Fundamentals of Enzyme Engineering

Young Je Yoo · Yan Feng · Yong Hwan Kim
Camila Flor J. Yagonia

Fundamentals of Enzyme Engineering

Springer

Young Je Yoo
School of Chemical and Biological Engineering
Seoul National University
Seoul, Korea

Yan Feng
College of Life Science and Biotechnology
Shanghai JiaoTong University
Shanghai, China

Yong Hwan Kim
School of Energy and Chemical Engineering
Ulsan National Institute of Science and Technology
Ulsan, Korea

Camila Flor J. Yagonia
School of Chemical Engineering
University of San Carlos
Cebu City, Philippines

ISBN 978-94-024-1469-1 ISBN 978-94-024-1026-6 (eBook)
DOI 10.1007/978-94-024-1026-6

Softcover reprint of the hardcover 1st edition 2017

Printed on acid-free paper

This Springer imprint is published by Springer Nature
The registered company is Springer Science+Business Media B.V.
The registered company address is: Van Godewijckstraat 30, 3311 GX Dordrecht, The Netherlands

Preface

Biotechnology has become one of the most important technologies in twenty-first century contributing every aspect of our life and industry. Biotechnology can contribute to the problems of our human culture and civilization from health, food, energy, environment, and to materials issues. Biotechnology is therefore playing a key role in pharmaceutical, medical, chemical, electronics, energy, and environment industries.

For the development of biotechnology, deep understanding and fusions in biology, chemistry, enzymology and engineering are required. One fundamental area of biotechnology is enzyme engineering which covers enzymology, enzyme technology, and engineering of enzymes. Enzymes have been used for food preparations such as cheese and alcohols from long time ago. In 1970s, immobilized enzymes have accelerated the development of enzyme engineering. In 1980s, the understanding of enzyme reaction in organic solvent has created a new area in enzyme engineering. Also with the energy and environment crisis, bio-based chemicals and bioenergy have opened a new area in enzyme engineering. With systems biology and metabolic engineering, enzymes for bio-based chemicals including polymers have become more and more important nowadays. Recently with the knowledge on molecular dynamics and quantum mechanics, computational tools have become stronger, which will contribute to the design of novel enzyme in the long run. Cheaper enzymes and more stable enzymes as well as more applications are required for their wide commercial use.

Even though enzymes are becoming more and more important, it is not easy to find a good textbook for the students to study the role of enzymes in biotechnology area, except for handbook style books or books dealing with current issues and specific topics, which gave us an idea to start to write the book, Fundamentals of Enzyme Engineering.

This book is written mainly for senior level or graduate students in biotechnology. However, this book can be also used as a guidebook for an overview of enzyme engineering working in relevant industries. This book consists of four parts with 15 chapters and deals with fundamentals of enzyme chemistry, classical enzyme reaction engineering, recent molecular level understanding of enzyme, and various applications of enzymes.

Even though there are so many research results and industrial experiences reported so far, fundamentals and basic concepts with some cases are introduced and emphasized in the text instead of a knowledge-oriented description of every case of enzyme engineering. For the details or for specific cases, reading and discussion using related references are desirable. Some of which are introduced in the text as case studies and examples. For industry, searching for the patents and discussion based on the patents are also desirable for understanding of the technology and for further development. Since our knowledge and understanding of enzymes is still not enough and many challenges are waiting, further discussions on these issues are therefore presented at the end of each chapter.

For students who are familiar with basic biotechnology including biochemistry and biology, it is recommended to emphasize advanced level and recent advancement of enzyme engineering including molecular understanding and applications which are the integration of diverse principles. Reactor design and optimization which is generally required in industry are not described in detail in this text, which can be supported using the texts on reactor design.

We would like to express our gratitude to many colleagues, friends, students who gave tips, comments, and assistance during the development of this book as well as to the publisher who gave us a chance to publish this book.

We hope this book Fundamentals of Enzyme Engineering can be useful for the students in academia as well as the engineers in industry for the future development of enzyme engineering and biotechnology.

Seoul, Korea	Young Je Yoo
Shanghai, China	Yan Feng
Ulsan, Korea	Yong Hwan Kim
Cebu City, Philippines	Camila Flor J. Yagonia

Contents

Part III Molecular Understanding of Enzymes

Part IV Applications of Enzymes

Part I
Introduction to Enzyme Engineering

Chapter 1
Introduction

1.1 Brief History of Enzyme Engineering

The microorganisms were widely used among ancient people. The manufacture of cheeses, breads, alcoholic beverages, and many other applications depends upon microorganisms which were found from ancient text of Babylon, Greece, Egypt, China, and India. Enzymes were the main components in microorganisms for the food manufacturing and other applications. However, ancient applications were relied upon observations and repeated experiences rather than scientific and technological. From the early of nineteenth centuries, enzymes have been investigated by scientists in a more systematic way. Tables 1.1 and 1.2 show the milestone and some Nobel Prize winners during this period. Many books relating to enzyme technology and engineering have been introduced (Wang *et al.* 1979; Dordick 1991; Fersht 1999; Bommarius and Riebel 2000) which can be good references in this area.

1.1.1 Brief History

The steroid biotransformations in the 1950s replaced the chemical reactions required to introduce a hydroxyl group at the 11α position and this marks the first bioconversion of industrial importance. Peterson *et al.* (1952) described this steroid modification using *Rhizopus nigricans*. Two other steroid modifications were also reported about the same time using *Curvularia lunata* and *Corynebacterium simplex* (Shull *et al.* 1953; Nobile *et al.* 1955).

There have been historically four steps of technology advances which brought a big impact for enzyme as alternative catalysts on bioconversions namely: enzyme isolation, enzyme immobilization, enzymes in non-conventional media, and recombinant DNA technology (Lilly 1994).

Y.J. Yoo *et al.*, *Fundamentals of Enzyme Engineering*,
DOI 10.1007/978-94-024-1026-6_1

Table 1.1 History and trends of enzyme engineering

Year	Significant event/discovery
1890	"Lock-and-Key" model was proposed (Fisher)
1893	Definition of the term "catalyst" (Ostwald) was introduced
1897	Explained that enzymes do not require a cell (Buchner)
1926	Enzyme was proved to be a protein (Sumner)
1952	Steroid biotransformation was performed
1958	"Induced-fit" model was proposed (Koshland)
1963	Amino acid sequence of ribonuclease was identified
1965	"Allosteric model" of enzyme was proposed (Monod)
1970	Immobilized enzymes were used for the production of HFCS
1980	Enzymatic synthesis of chiral compounds and polymers in organic solvent were reported
1990	Protein engineering including directed evolution was used
2000	Computational design of enzyme was introduced
2010	Enzymes for metabolic engineering become popular

Table 1.2 Nobel prize winners on enzymes/proteins

Year	Prize winner	Scientific discovery
1907	Eduard Buchner	Alcoholic respiration with cell-free extract
1909	Wilhelm Ostwald	Definition of the word "catalyst"
1946	James B. Sumner	Urease enzyme crystallized from jack beans
1958	Frederick Sanger	Structure of insulin
1972	Stanford Moore and William Stein	Connection between chemical structure and catalytic activity of ribonuclease
1993	Michael Smith	Site-directed mutagenesis for enzyme sequence change
	Kary B. Mullis	Invention of polymerase chain reaction (PCR)
2002	John B. Fenn, Koichi Tanaka and Kurt Wüthrich	Tools (Mass Spec) for identification and structure analysis of biomacromolecules
2008	Martin Chalfie, Roger Y. Tsien and Osamu Shimomura	Green fluorescent protein (GFP)
2013	Martin Karplus, Michael Levitt and Arieh Warshel	Multiscale models for complex chemical systems (Quantum mechanics/molecular mechanics)

The technology development for the separation of enzymes from microorganisms is one of the important technological advancements. The disruption of microorganisms by mechanical means (e.g., high-pressure homogenizer) allowed the isolation of intracellular enzyme. However, isolation of these intracellular enzymes was quite expensive; thus their applications had been limited. Engineering them for repeated use, higher activity and stability has allowed more feasible processes.

Table 1.3 Industrial bioconversions of substrates which are poorly soluble in water

Process	Enzyme	Company	Operating since
Fat interesterification	Lipase	Fuji Oil	1979
Ester hydrolysis	Lipase	Sumitomo	1988
Transesterification	Lipase	Unilever	1990
Aspartame synthesis	Thermolysin	DSM	1992
Acylation	Lipase	BASF	1996

Enzyme immobilization is a technique that converts a water-soluble enzyme into an insoluble form which can be easily recovered and reused. If properly designed, immobilization can enhance the applications of enzyme in many ways such as production of chemicals and pharmaceuticals, enzyme biosensor, etc. In the 1960s, several industrial technologies have been developed using immobilized enzymes, one example is the enzymatic isomerization of glucose to fructose for the production of high fructose corn syrup (HFCS). Clinton Corn Processing initiated the commercial production of HFCS in 1974 using glucose isomerase immobilized on an ion-exchange resin (Antrim *et al.* 1979). In 1976, the first continuous production process for HFCS was performed in Japan.

Generally, industrial enzymatic reactions before mid-1970s use substrates and products which are soluble in aqueous solutions. The production of cholestenone from cholesterol by *Nocardia* (NCIB 10554) in high proportions of water-immiscible solvent performed by Buckland *et al.* (1975) led to the third technological advance. Klibanov (1986) showed that enzymes can also function in organic solvents. Since then, various enzymatic processes using organic solvents have been industrially set up (Table 1.3).

The next technological advancement is recombinant DNA technology. This technology together with protein engineering allows new and better enzyme variants to be quickly produced. Since around 1990, directed evolution has been used as a powerful tool for enzyme engineering. This method contributed a lot in the development of enzymatic processes for industrial applications to harness the capability of naturally occurring enzyme since in most cases, natural enzymes are not optimized for industrial reaction conditions.

Current issues and recent advances in enzyme engineering is the computational design. The computational design of enzyme is another method that has been developed around year 2000. This is accomplished using computer models to suggest sequences and structures that can work for the desired properties of the enzyme. Understanding the mechanism of enzymes in detail and the structure of functional enzyme can make enzyme technology jump one more step. At present, the study of enzymes is still one of the important issues to the scientific community and to the industry sector in general. Artificial enzymes, catalytic antibody are examples of current issues in enzyme engineering. Recently, synthesis of ammonia through enzymes was reported (Brown *et al.* 2016), which can be a breakthrough of enzyme engineering in chemistry and chemical industry. Enzymes are continuously utilized for many industrial applications including their recent usages in chemicals production as well as their traditional roles. Many challenging issues are still waiting.

1.1.2 Enzyme Technology and Engineering

Enzyme technology is fundamental in biotechnology that the European Federation of Biotechnology defined: "Biotechnology is the integration of natural sciences and engineering sciences in order to achieve the application of organisms, cells, parts thereof and molecular analogs for products and services. In enzyme technology, many products such as food, fine chemicals and pharmaceuticals have been and are being manufactured utilizing enzymes as biocatalysts. Aside from that, enzymes are applied for analytical and diagnostic purposes. They are also used in many fields including environmental remediation. In recent years, enzymes for chemical synthesis replacing petrochemicals and for carbon dioxide utilization are new topics in enzyme engineering society."

Modern enzyme technology was started when it was shown that sugars can be obtained from starch using an alcohol precipitate of malt extract. The compound in the precipitate which can yield dextrins from starch was later called diastase. By the mid-nineteenth century, more enzymes were discovered including pepsin, invertase, and peroxidase.

After enzyme technology became established, enzymes as catalyst for industrial use were widely investigated. Taka-Diastase was patented for industrial application: amylolytic enzyme produced by *Aspergillus oryzae*. Now enzymes are being utilized for various applications from pharmaceuticals to diagnostics.

Biotechnological processes use one or more enzymes as biocatalysts depending on the required process condition. Compared to fermentation processes, enzymatic catalysis with isolated or immobilized enzymes has the following advantages: (1) by-product formation is minimized by other enzymes in the cells, (2) no need for complex nutrients medium (e.g., carbon, nitrogen, and other nutrient sources essential for cell growth), and (3) smaller reactors can be used since higher productivity can be obtained than with living cells. However, fermentation using living cells are still suitable than isolated enzymes when several enzymes in series and cofactor regeneration are involved.

Enzymes are more selective than conventional chemical catalyst. High selectivity is one of the main advantages including reduced side reactions and thus easier separations. In addition, enzymatic catalysis can be carried out under mild reaction conditions and it has, in many cases, high turnover numbers compared to the conventional chemical catalysis. Chemical catalysis is typically operated at much higher temperature and pressure.

However, there are also several issues associated with enzymatic catalysis for chemical synthesis: one of these is the difficulty for enzyme isolation. Enzymes

Table 1.4 Advantages and disadvantages of enzyme technology compared to chemical processes

Advantages	Disadvantages
High degree of selectivity Reactions at mild conditions	Expensive
Environmentally friendly	Unstable
Catalyze broad spectrum of reactions	Low productivities
Less by-products	
Non-toxic and non-flammable	

are inhibited by relatively low concentrations of end products, high temperature, extreme pH conditions, some metal ions, and solvents. In addition, some enzymes are still very expensive for industrial use and may require expensive cofactors for catalysis. Enzymatic catalysis tends to be too specific for general applicability. Table 1.4 summarizes the pros and cons of enzyme technology. Enzyme engineering is also widely used terminology covering traditional enzyme-related technology and engineering of enzymes which is very critically important nowadays to enhance the applicability of enzymes.

1.1.3 Classification of Enzymes

Approximately 5000 enzymes have been characterized so far, while more than 300 enzymes are commercially available and supplied from enzyme manufactures. Depending on the reactions they catalyze, enzymes are grouped according to the report of the Nomenclature Committee of the International Union of Biochemistry (1984). The six distinct classes are shown in Table 1.5.

Enzymes are named by adding the suffix—*ase* to the name of their substrate. However, there are also enzymes that have been given names that do not denote their substrates such as pepsin and trypsin. To avoid ambiguities, International Union of Biochemistry (IUB) assigned each enzyme a name and a four-level number. The Enzyme Commission (EC) numbers divide enzymes into six main groups depending on the reactions they catalyze as shown in Table 1.5. For this EC number system, the first, second, third, and fourth number refers to the class of enzyme, subclass by the type of substrate or the bond cleaved, subclass by the electron acceptor of the type of group removed and serial number of enzyme found, respectively. For example, as shown in Fig. 1.1, glucose isomerase is EC 5.3.1.5, also called xylose isomerase.

Some chemical reactions that are also catalyzed by enzymes are shown in Table 1.6. Alcohol dehydrogenase is also called aldehyde reductase. This enzyme acts on primary or secondary alcohols or hemi-acetals. Lipases can catalyze transesterification and other reactions in organic solvent system including esterification, amino lysis, acyl exchange, thiotransesterification, and oximolysis.

Table 1.5 Classification of enzymes

No.	Class	Representative subclasses	Type of reaction	Example
1	Oxido-reductases	Oxidases, oxygenases, peroxidases, dehydrogenases	Transfer of electrons (oxidation and reduction)	$-CH_2OH \leftrightarrow -CH{=}O$ $-SH \leftrightarrow -S-S-$ $-CH_2NO_2 \leftrightarrow -CH = O$ 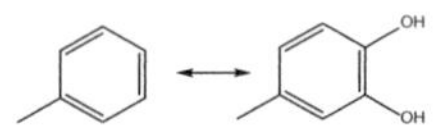
2	Transferases	Glycosyltransferases, methyltransferases, transaldolases, transketolases, acultransferases, alkyltransferases, transaminases, sulfotransferases, phosphotransferases, nucleotidyltransferases	Group-transfer reaction	$-CH_3$ $-CH_2OH$ $-CHO$
3	Hydrolases	Esterases, lipases, glycosidases, proteases, sulfatases, phosphalases, aminoacylases, endo- and exo-nucleases, halohydrolases	Hydrolysis reactions	$-COOR \leftrightarrow$ $-COOH + ROH$ $-COSR \leftrightarrow$ $-COOH + RSH$ $-CONH_2 \leftrightarrow$ $-COOH + NH_3$
4	Lyases	Decarbosylases, aldolases, ketolases, hydratases, dehydratases, polysaccharide lyases, ammonia lyases	Addition of groups to double bonds	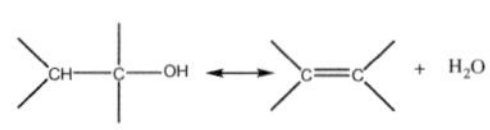
5	Isomerases	Racemases, epimerases, isomerases	Transfer of groups within molecules to isomeric forms	Glucose $\leftrightarrow$ fructose 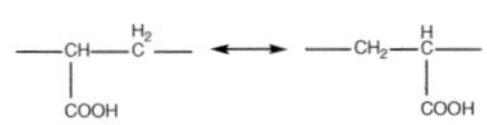
6	Ligases	Synthetases, carboxylases	Formation of C–C, C–S, C–O and C–N bonds	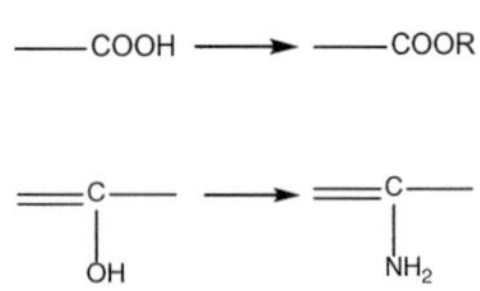

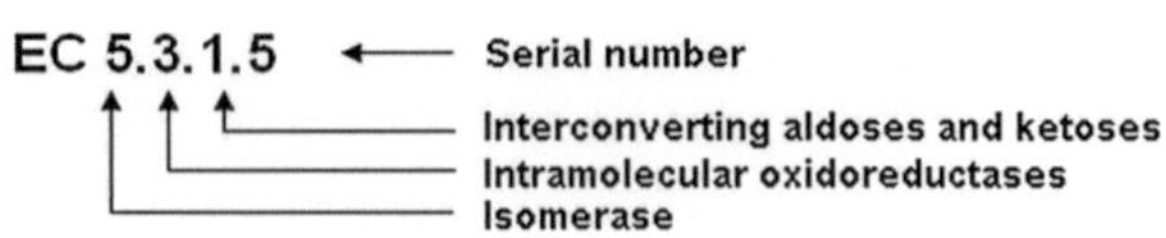

Fig. 1.1 EC number system for glucose isomerase

Table 1.6 Examples of enzyme-catalyzed reactions

Reaction	EC number	Enzyme
Meerwein-Ponndorf-Verley reduction	1.1.1.1	Alcohol dehydrogenase
Baeyer-Villiger oxidation	1.14.13.22	BV monooxidase
Ether cleavage	1.14.16.5	Glyceryl etherase
Disproportionation	1.15.1.1	Superoxide dismutase
Etherification	2.1.1.6	COMT[a]
Transamination	2.6.1.x	Transaminase
Oximolysis	3.1.1.3	Lipase
Aldol reaction	4.1.2.x	Aldolase
Racemization	5.1.2.2	Mandelate racemase
Claisen rearrangement	5.4.99.5	Chorismate mutase

[a]*COMT* Catechol-*o*-methyltransferase

The quantification of enzymes is often difficult to determine in absolute terms such as grams, since the activity changes due to conformations and the environments such as temperature and pH. More relevant parameter is to express the *enzyme activity* in terms of the activity unit (U), which is defined as the amount which will catalyze the transformation of 1 μmole of the substrate per minute under standard conditions or optimum pH and temperature and in the presence of specific chemicals if required. Another parameter of interest is the specific activity (e.g., U kg^{-1}) having some utility as an index of the enzyme purity.

1.2 Industrial Application of Enzymes

Enzymes have numerous applications in food, medical, chemical, and pharmaceutical industries. The industries have grown rapidly over the past decades and are expected to continue their growth. Table 1.7 shows the applications of biocatalysts and its production scale. For this enzymatic processes, various enzymes have been applied.

Table 1.7 Industrial application of enzymes and its production scale

Production scale	Product	Enzyme	Company
>1,000,000	High-fructose corn syrup (HFSC)	Glucose isomerase	Various
>100,000	Lactose-free milk	Lactase	Various
>10,000	Acrylamide	Nitrilase	Nitto Co.
	Cocoa butter	Lipase (CRL)	Fuji oil
>1000	Aspartame ®	Thermolysin	Tosoh/DSM
	Nicotinamide	Nitrilase	Lonza
>100	Ampicillin	Penicillin amidase	DSM-Gist Brocades
	(S)-methoxyisopropylamine	Lipase	BASF

Biocatalysts in industries are generally used to produce their natural products and derivatives. Carbohydrates and fatty acid derivatives are mostly used in food industry, while other types of compounds are mostly applied in pharmaceutical and agro industries. Pharmaceutical sector especially dominates applications of biocatalysts (Straanthof *et al.* 2002).

As discussed above, enzymes are being applied in various fields from food, pharmaceuticals, chemicals, and other industrial applications. At present, more than 5000 enzymes are known. Approximately more than 200 microbial-origin enzymes are used commercially. Commercial enzyme production has grown during the past decades in response to increasing demands and application for enzymes. Table 1.8 shows the leading enzyme manufacturers. The enzyme manufacture is relatively concentrated on a few countries such as Denmark, Switzerland, Germany, Netherlands, USA, Japan, Russia, and Korea.

For more applications

Enzymes have been used from old days which resulted in more understanding in enzymes, increasing demand and applications. However, many researchers in academia and industry are still looking for more applications and better technologies. The relationship between structure and function has been extensively investigated, but still remains as one of the hottest current issues in enzyme engineering. Recently, CRISPER (Clustered Regularly Interspaced Short Palindromic Repeats)-Cas system is widely investigated as a novel and powerful tool for studying gene regulation, gene expression, genome-wide screening after the introduction of DNA lyases long time ago and requires molecular understanding of the system to improve the specificity and other properties for further applications. With these backgrounds, activity and stability versus structure issues can find a solution for novel functions and applications. Since metabolic engineering can also contribute to the bio-based chemicals synthesis, enzyme engineering, and technology will play an important role as a key technology in metabolic engineering and systems biology. Also novel applications of current enzymes are also being considered as important issue. For example, in cosmetics industry, more natural and nontoxic ingredients are especially required, which can be obtained from natural products or enzymatic synthesis instead of chemical synthesis where by-products are in many cases formed.

Case Study: Enzymes in detergent industry

Enzymes have been used in detergent formulations since 1960s to overcome the eutrophication of water caused by phosphorus detergents. Proteases, amylases, lipases, and cellulases have been used to degrade protein, carbohydrate, and lipid stains on clothes. Using enzymes in detergent industry, energy can be saved by washing at lower temperatures with comparable efficiency with that at high temperature and using traditional surfactants. In summary, enzymes in detergent industry have many advantages as follows:

- Lower cost since used at low detergent concentrations,
- Acceptable to environment: biodegradability and no harmful impact on sewage treatment processes;

Table 1.8 Leading enzyme manufacturers

Company	Location	Established year	Major products
Novozymes	Bagsvaerd, Denmark	1921	Household care, food and beverage, bioenergy, feed and biopharmaceuticals
Dupont (Genencor)	Delaware, USA	1982	Biofuels, food ingredients, animal nutrition, textiles and detergent
DSM	Delft, the Netherlands	1952	Animal nutrition, food ingredients, personal care, pharmaceutical
Roche	Grenzacherstrabe, Switzerland	1896	Diagnostics, pharmaceuticals
Amano	Nagoya, Japan	1899	Pharmaceuticals, dietary supplement, biotransformation, diagnostics, food processing
BASF	Luwigshafen, Germany	1865	Feed additives, pharmaceuticals, detergents
KAO	Tokyo, Japan	1882	Beauty care, health care, home care
AB Enzymes	Feldbergstrasse, Germany	1907	Feed additives, food, textile, detergent, pulp and paper, biofuels
Verenium	San Diego, USA	2007	Animal health and nutrition, grain processing, oilfield services
Iogen	Ontario, Canada	1970s	Biofuels, pulp and paper, textile, grain processing and brewing, animal feed
Dyadic	Florida, USA	1979	Food, brewing and animal feed enzymes, biofuels, pulp and paper, textile enzymes
Meiji	Tokyo, Japan	1916	Food
Enmex	Tlalnepantla, Mexico	1961	Alpha-amylase, alkaline protease
Nagase	Osaka, Japan	1832	Pharmaceuticals, food, agriculture, household, textiles
Amicogen	Jinju, Korea	2003	Functional food ingredients
InnoTech MSU	Moscow, Russia	2009	Peroxidases, formate dehydrogenase, D-amino acid oxidase
SibEnzyme	Novosibirsk, Russia	1991	Restriction enzymes, ligases, polymerases

- Higher efficiency in stain removal,
- Less use of pollutants such as phosphate, bleach, and caustic.

The enzymes related to detergent have been traditionally isolated from nature, but nowadays are being engineered to provide better properties to meet the formulation conditions and conditions of washing processes where high temperature and high pH conditions are sometimes required.

Further Discussion

1. What are the motivation and limit for the enzyme to be used in industry? How we can overcome the disadvantages of enzyme reaction compared to chemical catalysis and fermentation process?
2. Search the backgrounds and history of novel enzyme discovery such as enzyme for PCR and novel applications of the enzyme such as glucose isomerase.
3. What enzymes are being produced in leading enzyme producers? What are the main application areas for major enzymes produced by leading enzyme producers?
4. What properties are required for the enzyme to be used as detergent? Washings are being performed at low temperature or mild temperature depending upon the country and the pH for washing is not neutral.

References

Antrim RL, Colilla W and Schnyder BJ. Glucose isomerase production of high fructose syrups. *Applied Biochemisty and Bioengineering*, 1979, 2:97–155.

Bommarius AS and Riebel BR. *Biocatalysis*, Wiley-VCH, 2000.

Brown KA, Harris DF, Wilker MB, Ramussen A, Khandka N, Hamby H, Keable S, Dukovic G, Peters JW, Seefeldt LC and King PW. Light-driven dinitrogen reduction catalyzed by a CdS:nitrogenase MoFe protein hybrid. *Science*, 2016, 352:448–450.

Buckland BC, Dunnill P and Lilly MD. The enzymatic transformation of water-insoluble reactants in nonaqueous solvents. Conversion of cholesterol to chlolest4-ene-3-one by a *Nocardia* sp. *Biotechnology and Bioengineering*, 1975, 17:815–826.

Dordick JS (ed). *Biocatalysts for Industry*, Plenum Press, 1991.

Fersht A. *Structure and Mechniam in Protein Science*, WH Freeman, 1999.

Klibanov AM. Enzymes that work in organic solvents. *ChemTech*, 1986, 16:354–359.

Lilly MD. Advances in biotransformation processes. *Chemical Engineering Science*, 1994, 49(2):151–159.

Nobile A, Charney W, Perlman PL, Henong HL, Payne CC, Tully ME, Jevnik MA and Hershberg EG. Microbiological transformation of steroids. I. D-1,4-diene-3-ketosteroids. *Journal of the American Chemical Society*, 1955, 77:4184.

Peterson DH, Murray HC, Eppstein SH, Reineke LM, Weintaub A, Meister PD and Leigh HM. Microbiological transformations of steroids. I. Introduction of oxygen at carbon-I 1 of progesterone. *Journal of the American Chemical Society*, 1952, 74:5933–5936.

Shull GM, Kita DA and Davisson JW. U.S. Patent 2658 023, 1953.

Straathof AJJ, Panke S and Schmid A. The production of fine chemicals by biotransformations. *Current Opinion in Biotechnology,* 2002, 13:548–556.

Wang DIC, Cooney CL, Demain AL, Dunnill P, Humphrey AE and Lilly MD. *Fementation and Enzyme Technology*, John Wiley, 1979.

Chapter 2
Biosynthesis of Enzymes

2.1 Basic Enzyme Chemistry

2.1.1 Amino Acids

An amino acid is a molecule that has the following formula:

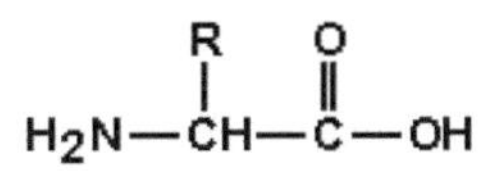

The central carbon atom covalently bonded by amino, carboxyl, and R group in the structure is called the *alpha carbon* (C_α).The side chain R group, vary in chemical composition, size, and interaction with water as reflected in their polarity. There are 20 standard amino acids used as common building blocks for peptides and proteins. The properties and structures of the side chains of these 20 naturally occurring amino acids are shown in Tables 2.1 and 2.2. Eight out of the 20 standard amino acids are called essential amino acids because they cannot be synthesized in our body and must be supplied from outside as food. They are isoleucine, leucine, lysine, methionine, phenylalanine, threonine, tryptophan, and valine.

Amino acids are amphoteric compounds. They have both a carboxylic acid group and an amino group; so they function as either an acid or a base depending on the pH of the environment. These two functional groups undergo an intramolecular acid base reaction to form a zwitterion (dipolar ion):

Y.J. Yoo *et al.*, *Fundamentals of Enzyme Engineering*,
DOI 10.1007/978-94-024-1026-6_2

Table 2.1 Name and properties of the 20 standard amino acids

Side chain	Name	Abbreviation	Molecular weight	pK_a of α-COOH	pK_a of α-NH_3^+	pKa of *R* group	pI
Nonpolar	Glycine	Gly or G	75.07	2.3	9.8		6.1
	Alanine	Ala or A	89.10	2.3	9.9		6.1
	Valine	Val or V	117.15	2.3	9.7		6.0
	Leucine	Leu or L	131.18	2.3	9.7		6.0
	Isoleucine	Ile or I	131.18	2.3	9.7		6.0
	Methionine	Met or M	149.21	2.1	9.3		5.7
	Proline	Pro or P	115.13	2.0	10.6		6.3
	Phenylalanine	Phe or F	165.19	2.2	9.3		5.7
	Tryptophan	Trp or W	204.23	2.5	9.4		5.9
Polar	Serine	Ser or S	105.10	2.2	9.2		5.7
	Threonine	Thr or T	119.12	2.1	9.1		5.6
	Asparagine	Asn or N	132.13	2.1	8.7		5.4
	Glutamine	Gln or Q	146.15	2.2	9.1		5.7
	Tyrosine	Tyr or Y	181.19	2.2	9.2	10.5	5.7
	Cystine	Cys or C	121.16	1.9	10.7	8.4	5.3
Acidic	Aspartic Acid	Asp or D	133.11	2.0	9.9	3.9	3.0
	Glutamic Acid	Glu or E	147.13	2.1	9.5	4.1	3.1
Basic	Lysine	Lys or K	146.19	2.2	9.1	10.5	9.8
	Arginine	Arg or R	174.20	1.8	9.0	12.5	10.8
	Histidine	His or H	155.16	1.8	9.3	6.0	7.6

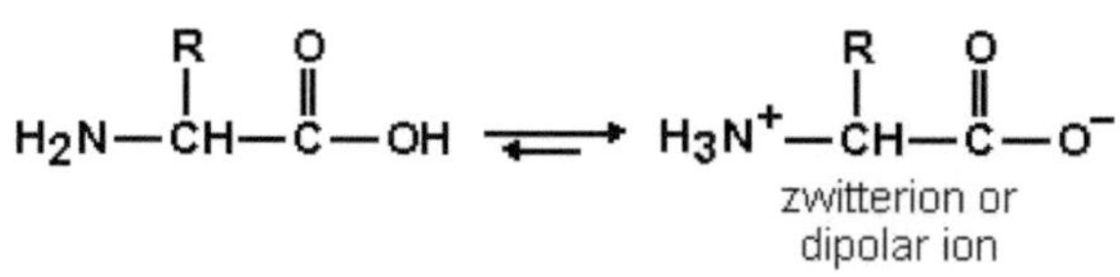

In the pH range near neutral the amino acid is in the dipolar ion form. In acidic solution, the carboxylate group becomes protonated and the amino acid is in its cationic form. At basic solution, the ammonium group gives up a proton and the amino acid exists as an anion:

Table 2.2 Structures of the 20 standard amino acids

Nonpolar Side Chains				
Glycine (G)	**Alanine (A)**	**Valine (V)**	**Leucine (L)**	**Isoleucine (I)**
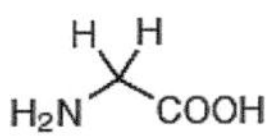	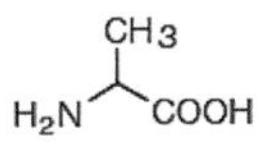	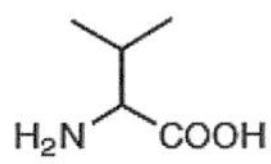	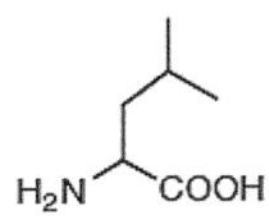	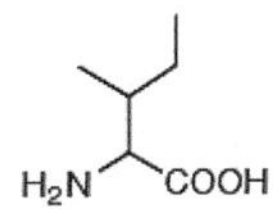
	Methionine (M)	**Proline (P)**	**Phenylalanine (F)**	**Tryptophan (W)**
	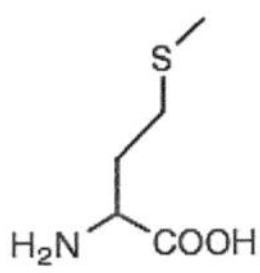	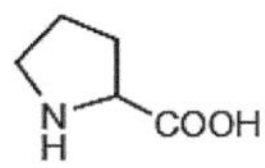	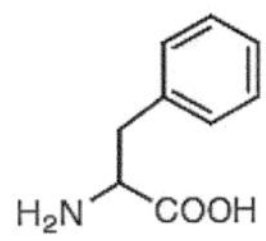	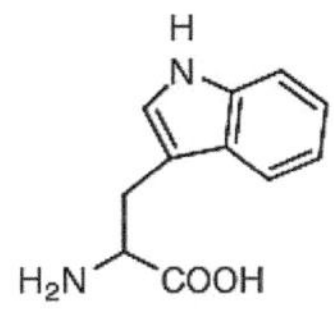

Polar Side Chains					
Serine (S)	**Threonine (T)**	**Asparagine (N)**	**Glutamine (Q)**	**Tyrosine (Y)**	**Cystine (C)**
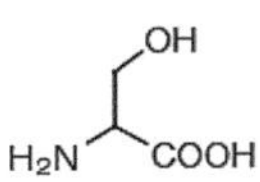	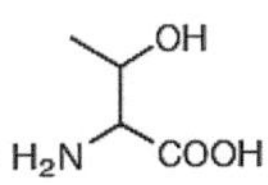	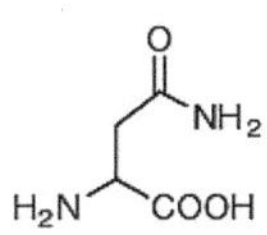	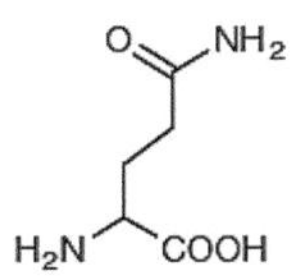	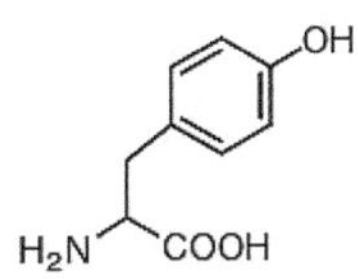	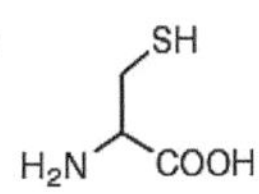

Electrically Charged Side Chains				
Acidic Side Chains		Basic Side Chains		
Aspartic Acid (D)	**Glutamic Acid (E)**	**Lysine (K)**	**Arginine (R)**	**Histidine (H)**
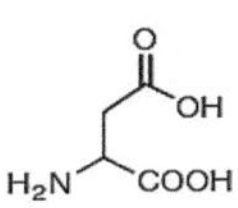	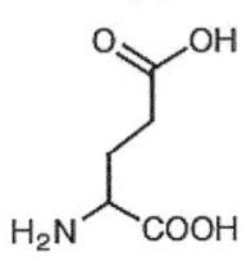	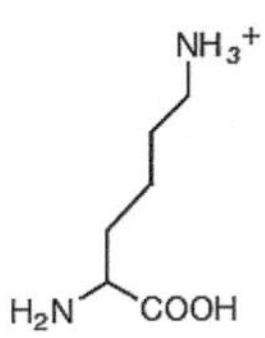	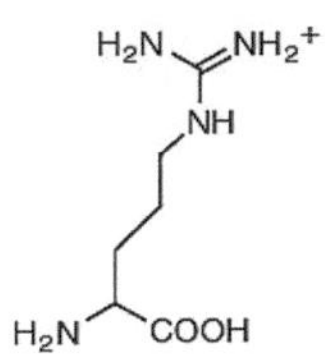	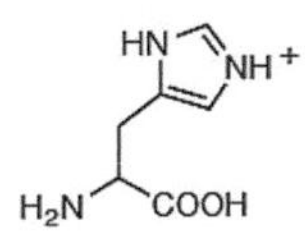

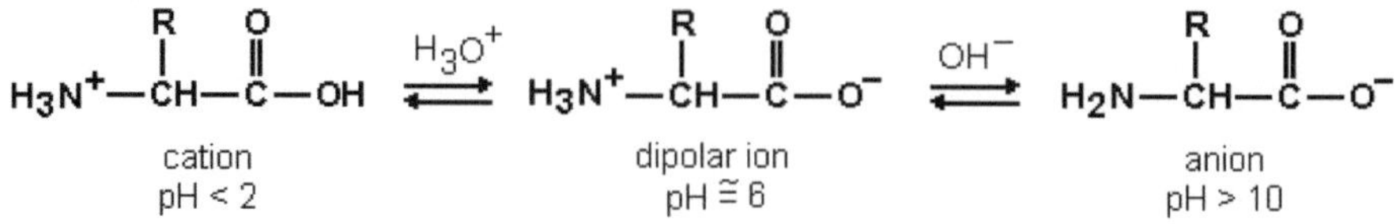

The pH at which an amino acid has overall neutral charge due to an equal proportion of negatively and positively charged groups is called the isoelectric point (pI).

2.1.2 Nonstandard Amino Acids

In addition to the 20 standard amino acids, selenocysteine and pyrrolysine are now regarded as the twenty-first and twenty-second amino acids. Selenocysteine (Sec/U) is a cysteine analog but a selenium-containing selenol group instead of a thiol group which provides unique properties, such as lower pK_a value (approximately 5.2) compared to that of cysteine (approximately 8.4) (Arnér 2010). On the other hand, pyrrolysine (Pyl/O) is an amino acid necessary for methanogenesis pathways (Krzycki 2013). Pyl contains a methylated pyrroline carboxylate in amide linkage to the ε-amino group of L-lysine (Hao *et al.* 2004).

Nonstandard amino acids are either found as minor components of some specialized type of proteins or through modifications of standard amino acids. 4-hydroxyproline and 5-hydroxylysine are among those that are derived from one of the 20 natural amino acids and both are found in the fibrous protein collagen. Homocysteine is formed through the transsulfuration from cystein (Brosnan and Brosnan 2006). *N*-methyllysine is found in myosin, a muscle protein.

There are also amino acids that occur biologically in either free or combined form. Examples include ornithine and citrulline which are derivatives of arginine and serve as intermediates in the formation of urea, part of amino acid catabolism (Curis *et al.* 2005). Aside from the derivatives of the α-amino acids found in proteins, some amino acids have their amino group in the β or γ position such as γ-aminobutyric acid (GABA) which serves as neurotransmitter and β-alanine which is an important precursor of the vitamin panthothenic acid. GABA is nowadays used as a starting substrates for bio-based chemicals. Examples of nonstandard amino acids are shown in Table 2.3.

Table 2.3 Structures of nonstandard amino acids

Selenocysteine	p-Acetylphenylalanine	p-Fluorophenylalanine	L-DOPA
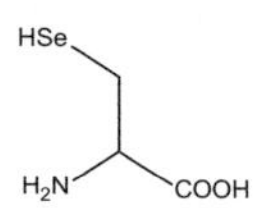	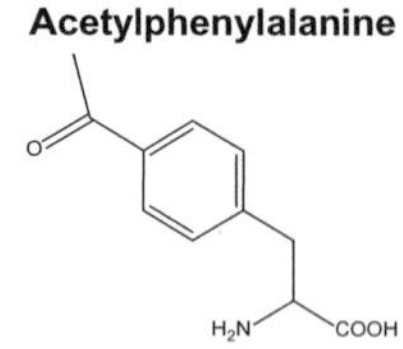	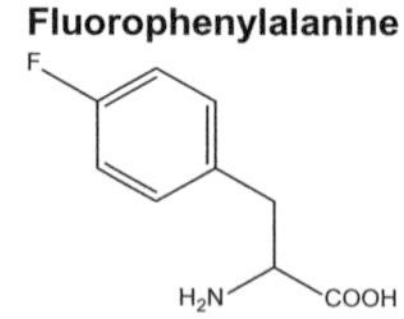	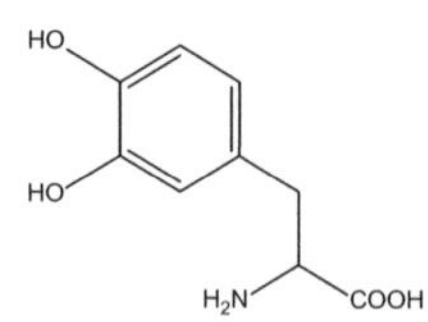

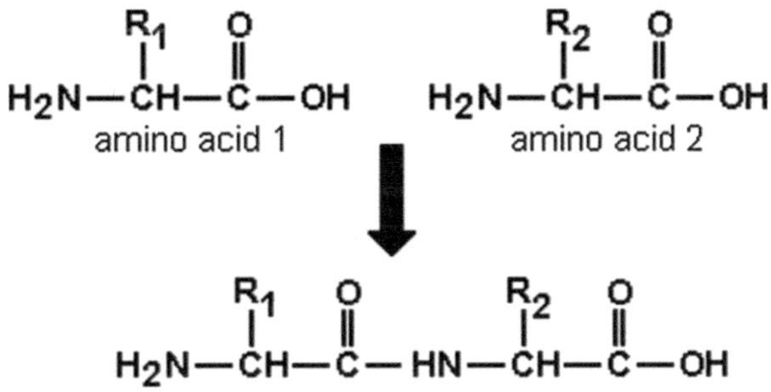

Fig. 2.1 Formation of peptide bond

2.1.3 Proteins

Amino acids can form amide bonds by condensation between carboxyl group and amino group as shown in Fig. 2.1. The amide bonds are specifically called the peptide bonds.

If two amino acids are condensed, the product is called as dipeptide. When another amino acid condenses to this dipeptide, a tripeptide is formed. In this manner, a chain of amino acids can be linked to make a *polypeptide* or a *protein*. On the basis of their physical characteristics, proteins can be classified into globular, fibrous, and membrane-bound classes. Globular proteins have polypeptide chains that are tightly folded into spherical shapes. They are soluble in aqueous systems and diffused readily. Fibrous proteins, on the other hand, are physically tough and water insoluble. The polypeptide chains in fibrous proteins are arranged in extended, parallel form along a single axis to give fiber structures. They function as structural or protective elements in the organism. The protein α-keratin for example is the major component of hair, feathers, nails, and skin. Another fibrous protein is collagen which is the major component of tendons.

Membrane proteins are also biologically important proteins. Within a cell and from cell to cell, membrane composition varies. Membrane proteins are found in a highly asymmetric environment which gives them unique properties. Outside the membrane is aqueous while the membrane interior is hydrophobic. Due to the two-dimensional surfaces of membrane proteins, they will concentrate or localized cellular components that regulates the nature and directionality of cell signals. However, their structural information has been known on a relatively small number of membrane proteins since isolating and crystallizing them is difficult.

There are four levels of protein structure (see Fig. 2.2) which are organized hierarchically from so-called primary structure to quaternary structure. The *primary structure* is the amino acid sequence of the polypeptide. The polypeptide chain that results through condensation reactions of amino acids retains a charged amino group at one end of the chain, N-terminus, and a charged carboxylate at the other end, C-terminus. The individual amino acids in a protein are numbered starting from the *N-terminus*.

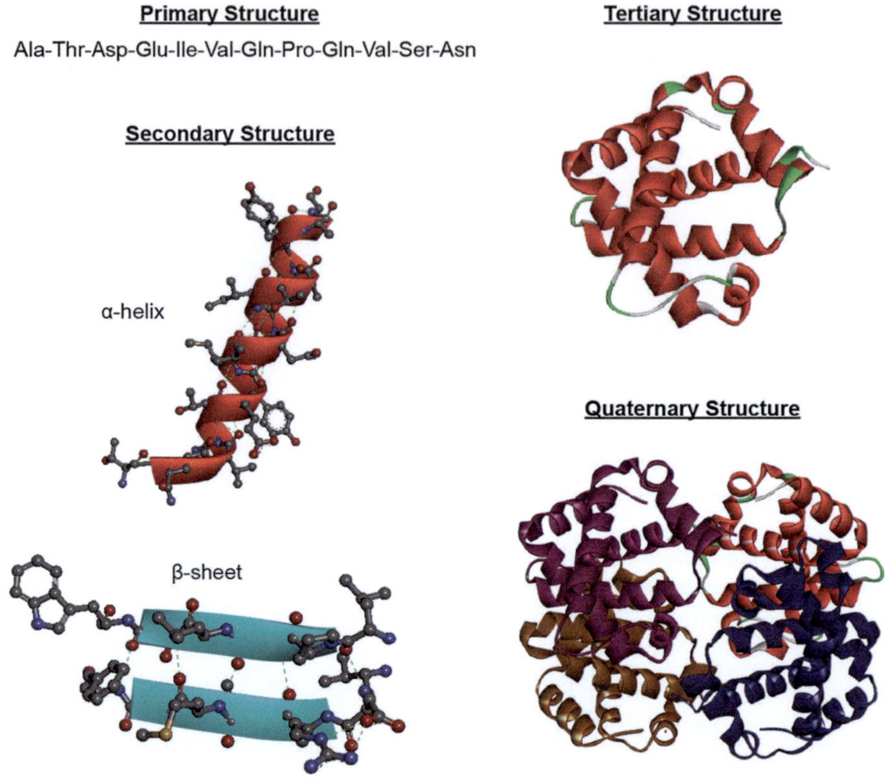

Fig. 2.2 Levels of protein structure

Secondary structure of protein refers to local sub-structures. *Alpha helix* and the *beta strand* (or *beta sheet)* are the two main types of secondary structure. These secondary structures are formed through specific patterns of hydrogen bonds between the main-chain amide and carboxylate groups.

Tertiary structure is a three-dimensionally folded structure due to secondary structure elements and interactions between side chains of the amino acids.

An assembly of several polypeptide chains into one large protein molecule is called the *quaternary structure*. Hemoglobin was the first oligomeric protein for which the complete tertiary and quaternary structures were determined.

2.2 Biosynthesis of Enzymes

2.2.1 Biosynthesis Mechanisms

Enzyme biosynthesis can be classified into two types: the *constitutive* and the *inducible*. Enzyme that is produced at a constant rate is called constitutive enzyme. In contrast, inducible enzyme is expressed only under specific conditions in which it can adapt to some environmental disturbance, thus its synthesis is not constant. One type of enzyme regulation is *induction* where enzyme expression is promoted by the presence of substrate or inducer. Contrary to the *induction* is *repression* where enzyme synthesis is suppressed by the presence of repressor molecules, for example intermediates or products of the enzyme reaction.

The foundation of enzyme production might be attributed to a genetic disturbance. The biosynthesis of an enzyme depends on (a) transcription of the genetic information from DNA or RNA into messenger RNA (mRNA) and (b) translation of the mRNA into polypeptide based on the genetic codon information in the mRNA. This is known as the operon model proposed by Jacob and Monod (1961).

The switching on and off for enzyme formation within a cell depends on the presence or absence of an effector molecule, which in turn is determined by the cell's environment. *Negative control* and *positive control* are the two main control mechanisms for the enzyme regulation. Four models of the switch are shown in Fig. 2.3 (Toda 1981).

For negative control of enzyme synthesis, the repressor (or its precursor) attaches to an operator region (*o*) in the DNA by itself (*R*) or as a complex (S_r-*R*) with an effector (S_r). This binding inhibits the transcription process, thus the reduced enzyme (*E*) synthesis.

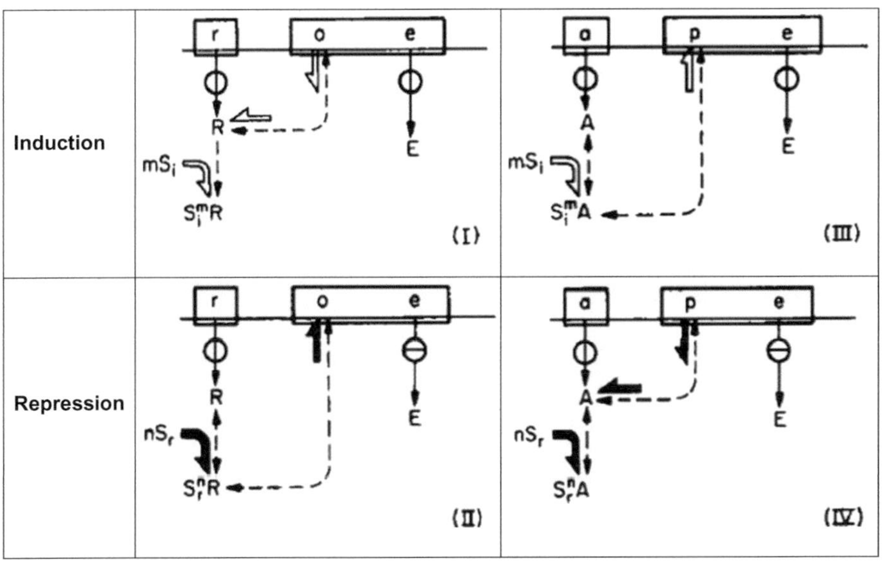

Fig. 2.3 Four types of enzyme regulation. *Dashed lines* indicate operator gene-repressor binding or promoter gene-activator binding. The *bold arrows* indicate the binding or the detachment of repressor or activator from the genes in the presence of an effector molecule. *A* activator; *a* gene of activator; *E* enzyme; *e* gene of enzyme; *m* and *n* stoichiometric constants; *p* promoter gene; *R* apo-repressor or repressor; *r* repressor gene; S_i inducer; S_r co-repressor (Toda 1981)

When an activator stimulates protein synthesis, this enzyme regulation is called positive control. The activator molecule (A) by itself or as a complex (S_i-A) with an effector (S_i) can bind on a promotor (p) in the DNA to enable transcription.

2.2.2 Mathematical Modeling

The enzyme regulation can be described by assuming the binding equilibrium between effectors (denoted as S_r for the co-repressor and S_i for the inducer), repressors (R and RS_r^n) and the free operator gene (O) for the induction system (Toda 1981):

$$R + mS_i \underset{K_1}{\Leftrightarrow} S_i^m R \quad K_1 = \frac{\left[S_i^m R\right]}{[R][S_i]^m} \tag{2.1}$$

$$R + O \underset{K_2}{\Leftrightarrow} RO \quad K_2 = \frac{[RO]}{[O][R]} \tag{2.2}$$

Total material balance for R and O gives

$$[R_t] = [R] + \left[S_i^m R\right] \tag{2.3}$$

$$[O_t] = [O] + [OR] \tag{2.4}$$

The enzyme biosynthesis rate (Q) or transcription rate is proportional to the ratio of free operator genes. For induction,

$$Q_i = \frac{[O]}{[O_t]} = \frac{1 + K_1[S_i]^m}{1 + K_1[S_i]^m + K_2[R_t]} \tag{2.5}$$

From Eq. 2.5 and for high concentration of inducer; $Q_i \cong 1.0$ which means full expression for enzyme biosynthesis.

However, for inducer concentration $S_i \cong 0$,

$$Q_i \cong \frac{1}{1 + K_2[R_t]} \neq 0 \tag{2.6}$$

This means minimal expression of enzymes.

For the repression system

$$R + nS_r \underset{K_1}{\Leftrightarrow} RS_r^n \quad K_1 = \frac{\left[RS_r^n\right]}{[R][S_r]^n} \tag{2.7}$$

$$RS_r^n + O \underset{K_2}{\Leftrightarrow} ORS_r^n \quad K_2 = \frac{\left[ORS_r^n\right]}{[O]\left[RS_r^n\right]} \tag{2.8}$$

$$Q_r = \frac{[O]}{[O_t]} = \frac{1 + K_1[S_r]^n}{1 + K_1[S_r]^n + K_1K_2[S_r]^n[R_t]} \tag{2.9}$$

Further Discussion

1. Explain how α-helix and β-sheet are formed. What are the roles of α-helix and β-sheet in enzyme structure and function? Helix might be used for flexible motion and sheet as rigid motion. Can we get new helix or sheet by changing one or more amino acids in helix or sheet, and how about the properties of the new helix or sheet?
2. How can nonstandard amino acids be incorporated into enzymes? What are the advantages that can be expected using nonstandard amino acids?
3. Induction and repression mechanism can be mathematically expressed in many ways. What are the advantages of expressing the mechanism mathematically and what applications?
4. Explain the induction mechanism and repression mechanism, respectively based on positive control mechanism.
5. Can we engineer inducible enzyme to constitutive enzyme? What could be the advantages and potential problems?

References

Arnér ESJ. Selenoproteins - what unique properties can arise with selenocysteine in place of cysteine? *Experimental Cell Research*, 2010, 316:1296–1303.

Brosnan J and Brosnan M. The sulfur-containing amino acids: an overview. *Journal of Nutrition*, 2006, 136 (6 Suppl): 1636S-1640S.21.

Curis E, Nicolis I, Moinard C, Osowska S, Zerrouk N, Bénazeth S and Cynober L. Almost all about citrulline in mammals. *Amino Acids*, 2005, 29(3):177–205.

Hao B, Zhao G, Kang P, Soares J, Ferguson T, Gallucci J, Krzycki J and Chan M. Reactivity and chemical synthesis of L-pyrrolysine — the 22nd genetically encoded amino acid. *Chemistry & Biology*, 2004, 11:1317–1324.

Jacob F and Monod J. Genetic regulatory mechanisms in the synthesis of proteins. *Journal of Molecular Biology*, 1961, 3:318–356.

Krzycki JA. The path of lysine to pyrrolysine. *Current Opinion in Chemical Biology*, 2013, 17:619–615.

Toda K. Induction and repression of enzymes in microbial culture. *Journal of Chemical Technology and Biotechnology,* 1981, 3:775–790.

Chapter 3
Production of Enzymes

3.1 Sources and Screening of Enzymes

3.1.1 Sources of Enzymes

The first step in enzyme production is the selection of the enzyme source. Enzymes can be derived from microorganisms through fermentation processes, as well as plant and animal sources. Table 3.1 presents industrially important enzymes and their sources.

Microorganisms are attractive sources of enzymes since they can be cultivated in a large scale (Fogarty 1983). Enzymes produced from microorganisms have, in many cases, better properties such as high stablity than enzymes from plant or animal sources. Identification of microorganisms that are suitable for the enzyme production starts by screening a wide variety of generally recognized as safe (GRAS) organisms if possible. On the other hand, screening of microorganisms considering their growth condition can also be a possibility. Extremophiles are microbes that can live and reproduce in harsh environments. Class of microorganisms in terms of growth conditions and some of the available enzymes in the microorganisms are shown in Table 3.2. Recent development in the applications of hyperthermophiles and their enzymes were reported (Atomi *et al.* 2011). There are many institutes where thermophilic and psychrophilic enzymes are being screened, and studied. One example is Exter Biocatalysis Center in University of Exter (United Kingdom).

Plants are not generally considered as sources of industrial enzymes since they are seasonal and the enzymes are intracellular, requiring additional processes to break the firm cell wall. However, despite these limitations, a number of enzymes from plant sources have been used in industry. For example, papain, cysteine protease can be obtained from the green fruit and leaves of the plant *Carica papaya*.

Y.J. Yoo *et al.*, *Fundamentals of Enzyme Engineering*,
DOI 10.1007/978-94-024-1026-6_3

Table 3.1 Industrially important enzymes and their sources

Enzyme	Source	Application
Derived from microbes		
Proteases	*Bacillus amyloliquefaciens, B. subtilis, Aspergillus oryzae, Streptomyces* spp.	Enzymatic hydrolysis of proteins
α-amylase	Various *bacilli* and *Aspergillus oryzae*	Hydrolysis of starch
β-amylase	Various *bacilli,* barley	Degradation of starch
Glucoamylase	*Aspergillus niger, Rhizopus species*	Starch hydrolyzed to glucose syrup.
Glucose isomerase	Various *bacilli, Streptomyces* spp.	Glucose isomerized to fructose
Invertase	*Saccharomyces* spp.	Sucrose hydrolyzed to glucose and fructose
Penicillin acylase	*E. coli*, various *bacilli, Streptomyces* spp.	Production of semisynthetic penicillins
Pectinases	*Aspergillus niger*	Enzymatic hydrolysis of pectin
Lipase	*Candida antarctica*	Hydrolysis of triglycerides
Derived from plants		
Papain	*Carica papaya*	Meat tenderization
Bromelain	Pineapple fruit/stem	Chill proofing of beers
Actinidin	Kiwi fruit	Meat tenderization
Lipoxygenase	Soybean	Bread making and aroma production
Derived from animal sources		
Chymosin (rennet)	Stomach of calves	Cheese manufacture
Ancrod	Snake venom	Anticoagulant
Urokinase	Urine	Thrombolytic
Trypsin	Animal pancreas	Digestive aid
Pancreatin (amylase, protease, lipase)	Pancreatic extract	Digestive aid
Pepsin	Stomach of animals	Digestive aid
Acetyl-cholinesterase	Bovine erythrocytes	Analysis of organophosphorus compounds such as pesticides
Cholesterol esterase	Porcine pancreas	Detect serum cholesterol levels

Table 3.2 Classification of microbes in terms of growth condition

Microorganisms	Growth condition	Enzymes and other biomolecules
Thermophiles	50–110 °C	Amylases, lipase, xylanases
Mesophiles	20–50 °C	Almost all enzymes
Psychrophiles	1–20 °C	Proteases, dehydrogenase
Alkaliphiles	pH > 9	Cellulases, proteases
Halophiles	3–20% salts	Compatible solutes, membranes

Animal tissues and animal secretes are also a potential source of enzymes. Rennin, also known as chymosin, is one of the industrially important animal derived enzymes. Rennin, aspartic protease, obtained from the stomach or abomasums of calves, is used for cheese production and as a digestive aid.

3.1.2 Screening of Enzymes

Enzymes are essential since their catalytic power facilitates life processes. However, several factors have to be considered such as enzyme substrate specificity, activity, selectivity, stability and recyclability to determine the usefulness of enzyme for biocatalysis. Enzyme activity, one of the most important properties, is affected by their structure and environment (see Fig. 3.1).

Nowadays, researches are geared toward modification of the enzyme to improve its catalytic potential for industrial use. Since the industrial application of enzymes depends on their activity, stability and cost; it is of great interest to engineer enzymes (e.g., screen novel enzymes and find new functions of the enzymes) that satisfy producers and end users criteria. One of the most recent and efficient strategies to discover novel enzymes is to find microorganisms with their characteristic diversity. Another strategy is the metagenome-based approach, the culture-independent method using microbial genomic data (metagenome). The process consists of four main steps:

1. obtaining samples from environment,
2. isolation of DNA and manipulation of the genetic material,
3. construction of metagenomic library, and
4. screening of new function enzyme and sequencing of genetic material from the metagenomics library (Lorenz and Eck 2005).

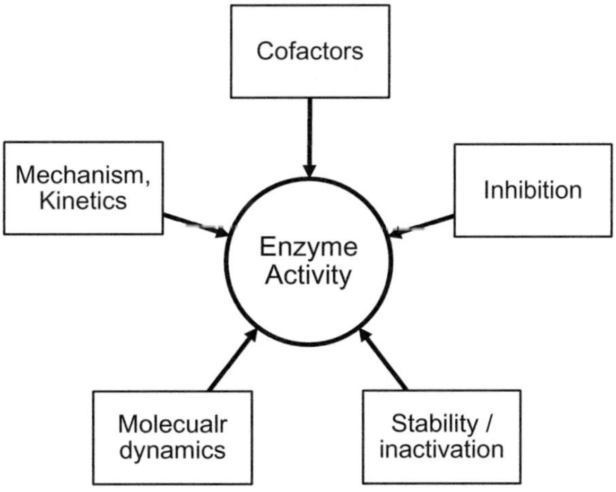

Fig. 3.1 Factors affecting enzyme activity

The followings are being considered for screening a microorganism for enzyme production: the microorganism should grow on cheap and readily available nutrients, the enzyme is to be produced with high yield in a relatively short time, and the microorganism should be nonpathogenic (Cheetham 1987). A number of techniques have been developed to obtain new enzymes. These include the following:

1. engineering at genetic level to mutate preexisting enzyme,
2. combining chemical and enzyme catalysts,
3. applying existing enzymes to novel unnatural substrate, and
4. use of different reaction conditions, such as solvent.

Examples of screening of novel enzymes

Traditional screening of novel enzyme is labor-intensive and time consuming. It was known that cephalosporin-synthesizing strain was screened from sewage in Sardinia, Italy. Acrylamide producing strain was screened from wastewater treatment site in acrylonitrile plant in Japan. Once, strain for specific purpose was screened, then enzymes in the strain can be obtained. Only 1% of the microorganism in the world has been used for screening, which means there are huge potential in screening to find novel enzymes. There are many institutes in the world handling and doing collections of enzymes including extreme enzymes.

Thermophilic enzymes can be obtained from the thermophilic microorganisms which can be found from hot spring, submarine volcanic vents, etc. For example, DNA polymerase enzyme for PCR was found from hot spring in Yellow Stone in USA in 1968 and volcano in Italy in 1985. To obtain thermophilic enzyme, first thing to do is to grow the microorganism at elevated temperature and screen the enzyme later from the survived microorganisms.

Benzene-degrading enzymes can be found by microbial enrichment technique using samples from oil-refinery or petrochemical plant site. By growing the microorganisms using soil samples from such sites by increasing gradually the benzene concentration, benzene-degrading enzymes can then be obtained.

New techniques have been suggested to find new enzymes. By changing enzyme structure, specifically by changing amino acid sequence and by employing different amino acids, nonnatural enzymes can be generated. This kind of protein engineering can be used to find better enzymes. At present, two different approaches have been widely used independently or in combination to improve the enzymes, changing amino acid sequences of enzymes: rational design and directed evolution. Rational design uses the knowledge of enzyme's structure-function relationship to modify the structure and improve the function. Directed evolution is similar to evolution and natural selection in the nature that random mutagenesis is introduced to an enzyme and the mutants with desirable properties are selected.

Sometimes, new unnatural substrates were found from existing enzymes: glucose isomerase was obtained from xylose isomerase, after examining glucose instead of xylose as substrate.

Once the enzyme reaction is identified at the early development stage for commercial application of enzyme, the first step is to find candidate enzymes using protein data bases. Then, increase the activity and stability using directed evolution technology and/or rational design technology, followed by large-scale production of the enzyme for real applications. The whole process in many cases takes around 6 months in industry.

3.2 Production of Enzymes

3.2.1 Media for Enzyme Production

There are various factors that influence a microorganism's metabolism and the enzymes produced in the pathway. The media in which the microorganism is cultured, for example, can affect the growth of microorganism and the production of enzymes. Culture conditions must be optimized for maximum production of microbial strains which in turn gives optimal production of the desired enzymes. Each microbial species grow at different rates with specificity to different substrate in the culture medium. These growth conditions can influence the enzymatic activities.

Alpha-amylases are extracellular enzymes which randomly cleave the α-1,4 glucosidic bonding of linear amylase and branching amylopectin. Bacterial α-amylases have been widely used in many industrial processes and can be produced by different microorganisms. An example of medium preparation for the bacterial amylase production is presented in Table 3.3 (Underkofler 1954).

Cellulase enzymes are becoming more important since they are recently applied for bioethanol production from lignocellulosic biomass. Cellulases can be produced using *Trichoderma reesi*. Ahamed and Vermette (2009) investigated the effects of fungal morphology on the cellulase productivity of *Trichoderma reesi* cultured in fed-batch bioreactor. The usage of a cellulose-yeast extract culture medium, as shown in Table 3.4 yielded the highest enzyme production.

Cheap carbon and nitrogen sources are very critical in industry to be competitive in the market, while defined media is important in academia to find bottleneck or limiting component. For example, glycerol which can be obtained as by-product

Table 3.3 Composition of medium for bacterial amylase production

Component	Composition (%)
Ground soybean meal	1.85
Autolyzed brewers yeast fraction	1.50
Distiller's dried soluble	0.76
Enzymatic casein hydrolysate	0.65
Lactose	4.75
$MgSO_4 \cdot 7H_2O$	0.04
Antifoam	0.05
Water	90.40

Table 3.4 Cellulose-yeast extract medium

Component	Composition (g/L)
Cellulose	10
Yeast extract	10
Glucose	10
$(NH_4)_2SO_4$	1.4
KH_2PO_4	2
$CaCl_2 \cdot 2H_2O$	0.4
$MgSO_4 \cdot 7H_2O$	0.3
$FeSO_4 \cdot 7H_2O$	0.005
$CoCl_2 \cdot 6H_2O$	0.0037
$MnSO_4 \cdot H_2O$	0.0016
$ZnSO_4 \cdot 7H_2O$	0.0014

from biodiesel industry can be used nowadays as cheap carbon source for microbial growth. Metabolic engineering should be thus incorporated if required for cheap media in industry.

3.2.2 Fermentation Process

Enzyme production is still an important field of biotechnology. Patents and research articles are increasing in numbers and the sales for enzymes would be close to a billion of dollars annually. Fermentation processes for microbial production of enzymes can be carried out through solid state culture or using submerged culture of microorganisms. Most enzyme manufacturers have produced enzymes using submerged fermentation (SMF) techniques. Submerged fermentation is defined as fermentation in the presence of excess water. This fermentation technique offers better monitoring and control over the process parameters, such as temperature, pH, aeration, and dispersion for efficient growth. Another advantage will be the ease of handling and possible scale-up, thus almost all the large-scale enzyme producing facilities are using this technology. However, the titers of this technology are relatively low. Product recovery cost is inversely proportional to product concentration in a fermentation broth, so the concentration in submerged fermentation must be optimized for a commercially feasible production of enzymes.

Solid state fermentation (SSF) where the microorganisms are grown on moistened solid substrate is still performed widely for cheap enzymes and where weather condition is good and labor is cheap. Microorganisms, such as fungi, yeast, some bacteria, or combinations of those can be used in SSF. The production of cheap glucoamylase and fungal spores for biocontrol can be produced by SSF (Ishida *et al.* 2000; de Vrije *et al.* 2001). SSF has many advantages over SMF in terms of simple operation, low capital investment, and low energy requirement. Agricultural wastes can be used as substrates without extensive pretreatment in SSF. However,

Table 3.5 Different culture methods

Solid state fermentation	Submerged fermentation
Requires much space for trays	Uses compact and closed fermenters
Requires much labor	Requires minimum of labor
Little power requirement	Needs power for air compressors and agitators
Minimum control is necessary	Requires control of culture condition
Not much separation for applications	Various separation technology is required

the development of SSF is slower than SMF since SSF has difficulties with process control and scale-up. Table 3.5 shows the difference between the two techniques.

3.2.3 Separation and Purification of Enzymes

As a preliminary step, the enzymes must be extracted from their biological sources. See Fig. 3.2 for the extraction procedures. Each extraction or purification procedure entails an increase in production cost and consequently lowers enzyme yield. Specific enzyme activity can be increased by adding purification stages which require also additional cost.

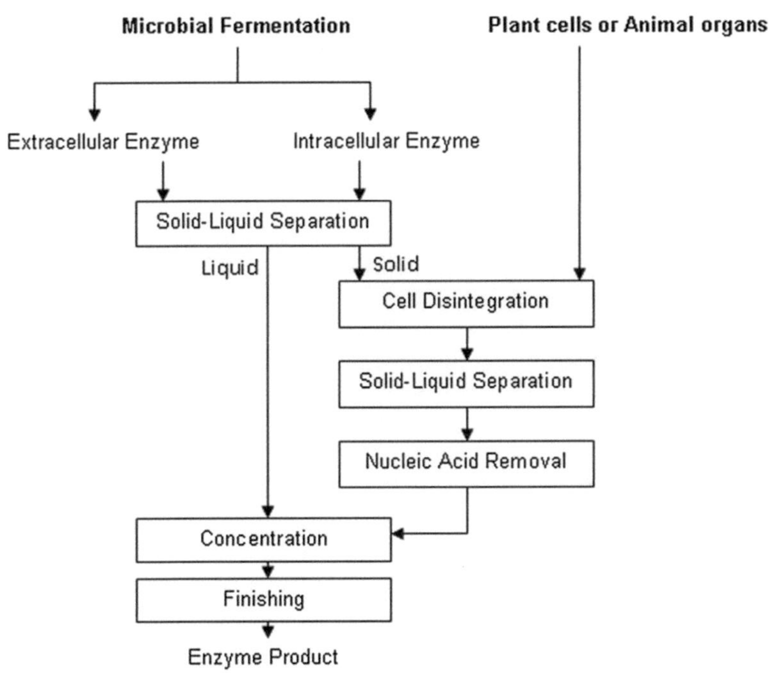

Fig. 3.2 Preparation of enzymes

From the biological source, the enzyme protein content should be high in order to ease the downstream processes. Another important factor to consider is enzyme activity. Maximum activity should be retained during the preparation of the enzymes. Heat, proteolysis, pH, oxidation and loss of cofactors are among the factors that lower the enzyme activity. The most significant thing among these factors is probably the heat inactivation especially if insufficient cooling is available during extraction and purification. Microbial contamination with proteases is also another reason for enzyme inactivation. Proteolysis by protease can be occured in the early stage of enzyme extraction since the proteases for enzyme degradation are still present.

Enzymes that are produced extracellularly are sometimes used without isolation and purification. On the other hand, cell disruption is required for the intracellular enzymes be released from the cells into the solution before purification. Solid-liquid separation is required for cell mass separation and later cell debris separation after breaking the cell. Cell disruption is needed to excrete intracellular enzymes. It involves giving shear forces to cells or adding chemicals for cell lysis. For shear forces, the degree of energy required to break the cells depends on the organism. Intracellular enzymes may contain nucleic acids, which can interfere enzyme purification procedures due to an increase in viscosity. Nucleic acids must be removed by the addition of nucleases or by precipitation. However, precipitating nucleic acid using ammonium sulfate can also remove some valuable protein at the same time. Ammonium sulfate as precipitating agent is widely used because of its high solubility, cheap price, lack of toxicity to enzymes. Enzyme preparations that have been concentrated can be further purified by chromatography. Table 3.6 shows the downstream processes for the separation and purification of enzymes.

Case Study: Expression and purification of Lipase A (Pfeffer *et al.* 2006)

Lipases from microorganisms have many industrial applications. Two lipases, lipase A and lipase B can be obtained from *Candida antarctica. Candida antarctica* lipase B (CALB) is well characterized. *Candida antarctica* lipase A (CALA) on the other hand has fewer available data but it has a big potential for industrial applications. CALA might be useful for the conversion of highly branched substrates that cannot be hydrolyzed by other lipases. The study was undertaken to characterize lipase A from *C. antarctica* which is expressed in the methylotrophic yeast *P. pastoris*. Figure 3.3 shows the methodology taken to obtain the desired enzyme product.

Two different fermentation methods were used: fed-batch and semi-continuous culture. The yield and other parameters are shown in Table 3.7. The protein concentration and the specific activity in the harvested culture supernatant were higher in the fed-batch process. However, the total activity of the purified CALA from the semi-continuous process was higher than in the fed-batch process. CALA has maximal activity at approximately 50 °C and about pH 7.0.

Table 3.6 Example of downstream processes for enzyme separation and purification

Methods	Description
Solid–liquid separation	
Centrifugation	Separates on the basis of the particle size and density difference between the liquid and solid phases Efficiency depends on the solids volume fraction, the effective clarifying surface
Filtration	Separates on the basis of particle size Efficiency is limited by the shape and compressibility of the particles, the viscosity of the liquid phase and the maximum allowable pressures
Cell disruption	
Ultrasonic cell disruption	Utilizes the sinusoidal movement of a probe with in the liquid Noise problem in lab and good for lab scale operation
Bead mills	Cell suspensions are agitated in the presence of glass or steel beads Cells are broken by the high shear gradients and collision with the beads
High-pressure homogenizer	Main disruptive factor is the pressure applied and consequent pressure drop across the valve
Lytic method	Enzymatic lysis has been used widely on the laboratory scale Lysis by acid, alkali, surfactant, and solvents can be effective in releasing enzymes, provided that the enzymes are sufficiently robust
Concentration	
Precipitation	Cheap method and first step in the purification of enzymes Increasing the ionic strength of the solution that causes a sufficient reduction in the repulsive effect of like charges between identical molecules of a protein and the forces holding the solvation shell around the protein for it to be precipitated
Ion-exchange chromatography	Positively charged (e.g., pH below isoelectric point) enzymes bind to cation exchangers Negatively charged (e.g., pH above isoelectric point) enzymes bind to anion exchangers The binding strength is determined by the pH and ionic strength of the solution and the structures of the enzyme and ion-exchanger
Affinity chromatography	Separates enzymes on the basis of a reversible interaction between the enzyme and a specific ligand immobilized to a chromatography matrix High selectivity and capacity for the protein of interest
Gel exclusion chromatography	Separates enzymes in the basis of their molecular size and shape The larger the molecule the more difficult for it to pass through and penetrate the beads

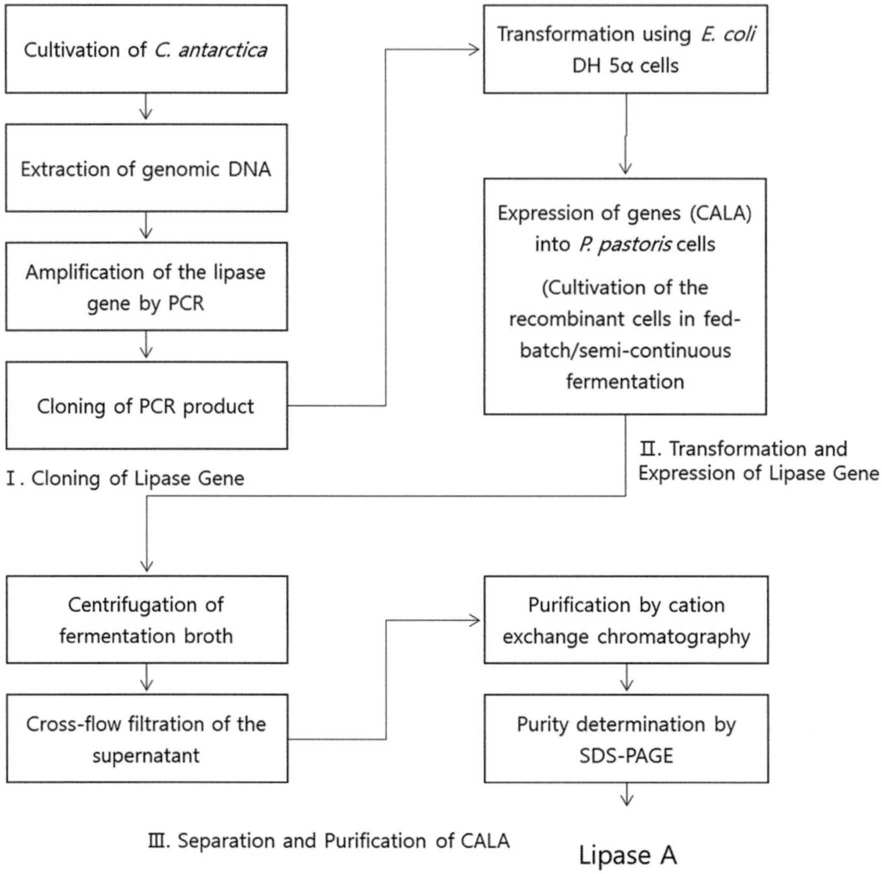

Fig. 3.3 Experimental methods for the expression and purification of CALA

Table 3.7 Overview of the results obtained with different lipase A expression strategies of and the subsequent purification of the enzyme (Pfeffer *et al.*, 2006)

Parameter	Shake flasks	Fed-batch fermentation	Semi-continuous fermentation
Processing time (days)	5	12	15
Protein concentration (supernatant) (g/L)	0.6	3.1	3.0
Specific activity (supernatant) (U/mg)	215	653	380
Specific activity (after purification) (U/mg)	–	2, 033	–
Total activity (U)	12,900	7, 530,000	10,233,000

Further Discussion

1. One of the popular enzyme is horseradish peroxidase (HRP) from plant. What kinds of plants? How to extract the enzyme and what applications?
2. Are the enzymes in thermophilic microorganisms also thermophilic?
3. What are the characteristics of thermophilic enzymes and psychrophilic enzymes compared to mesophilic enzymes? What principles can we apply to engineer mesophilic enzyme to thermophilic or psychrophilic enzyme?
4. How can we synthesize enzymes *in vitro* using the information on amino acid sequence?

References

Ahamed A and Vemette P. Effect of culture medium composition on *Trichoderma reesei*'s morphology and cellulose production. *Bioresource Technology*, 2009, 100:5979–5987.

Atomi H, Sato T and Kanai T. Application of hyperthermophiles and their enzymes. *Current Opinion in Biotechnology*, 2011, 22:618–626.

Cheetham PSJ. Screening for novel biocatalysts. *Enzyme Microb. Technol.*, 1987, 9:194–213.

Fogarty WM (ed). Microbial amylases. in *Microbial Enzymes and Biotechnology*. Applied Science Publishers, 1983, 1–92.

Ishida H, Hata Y, Kawato A, Abe Y, Suginami K and Imayasu S. Identification of functional elements that regulate the glucoamylase-encoding gene (glaB) expressed in solid-state culture of *Aspergillus oryzae*. *Current Genetics*, 2000, 37:373–9.

Lorenz P and Eck J. Metagenomics and industrial applications. *Nature Reviews Microbiology*, 2005, 3:510–516.

Pfeffer J, Richter S, Nieveler J, Hansen CE, Rhlid RB, Schmid RD and Rusnak M. High yield expression of Lipase A from in the methylotrophic yeast *Pichia pastoris* and its purification and characterization. *Applied Microbiology and Biotechnology*, 2006, 72: 931–938.

Underkofler LA. in *Industrial Fermentations* (Underkofler LA and Hickley RJ ed.), Chemical Publishing, 1954, 2:97–121.

de Vrije T, Antoine N, Buitelaar RM, Bruckner S, Dissevelt M, Durand A, Gerlagh M, Jones EE, Luth P, OOstra H, Rayensberg WJ, Renaud R, Rinzema A, Weber FJ and Whipps JM. The fungal biocontrol agent Coniothyrium minitans: production by solid-state fermentation, application and marketing. *Applied Microbiology and Biotechnology*, 2001, 56:58–68

Part II
Enzyme Reaction Engineering

Chapter 4
Enzyme Reaction Kinetics

4.1 Michaelis–Menten Kinetics

4.1.1 Kinetics

Since enzyme reaction, in many cases, follows first-order kinetics at low substrate concentration, and zero-order kinetics at high substrate concentration, simple enzyme reaction mechanism was suggested. During the course of enzyme, enzymes form a complex with the substrate. The mechanism is called as Michaelis–Menten kinetics for one-substrate reaction (Fig. 4.1).

The reaction sequence can be presented for one-substrate reaction as

$$\textbf{enzyme} + \textbf{substrate} \overset{K}{\leftrightarrow} (\textbf{enzyme} - \textbf{substrate complex}) \overset{k}{\rightarrow} \textbf{enzyme} + \textbf{product}$$

$$E + S \underset{k_{-1}}{\overset{k_1}{\longleftrightarrow}} ES \overset{k_2}{\rightarrow} E + P, \tag{4.1}$$

where k_1, k_{-1}, and k_2 are the respective rate constants.

The product formation rate, υ, is given as

$$\upsilon = \frac{\mathrm{d}[P]}{\mathrm{d}t} = k_2[ES]. \tag{4.2}$$

The rate of the enzyme–substrate complex [ES] formation is

$$\frac{\mathrm{d}[ES]}{\mathrm{d}t} = k_1[E][S] - (k_{-1} + k_2)[ES]. \tag{4.3}$$

Enzyme balance is

Y.J. Yoo *et al.*, *Fundamentals of Enzyme Engineering*,
DOI 10.1007/978-94-024-1026-6_4

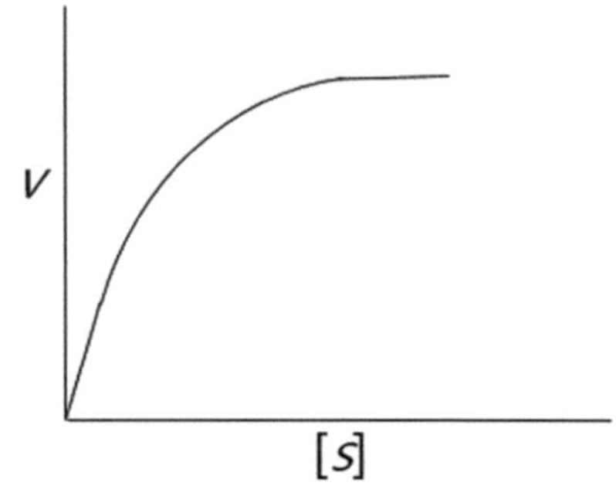

Fig. 4.1 Experimental observation of enzyme reaction rate

$$[E_0] = [E] + [ES]. \tag{4.4}$$

Thus

$$\frac{\mathrm{d}[ES]}{\mathrm{d}t} = k_1([E_0] - [ES])[S] - (k_{-1} + k_2)[ES]. \tag{4.5}$$

Assuming that the reaction is under quasi steady-state condition which means the rate of formation of ES equals its rate of disappearance by product formation and reverse reaction to substrate, such as

$$\mathrm{d}(ES)/\mathrm{d}t = 0. \tag{4.6}$$

From Eqs. 4.2 to 4.6, Michaelis–Menten equation is derived

$$\upsilon = \frac{k_2[E_0][S]}{\frac{k_2+k_{-1}}{k_1} + [S]} = \frac{V\mathrm{max}[S]}{K_\mathrm{m} + [S]}, \tag{4.7}$$

where K_m is the Michaelis constant and V_max is the maximum rate of reaction.

4.1.2 Physical Meaning

K_m corresponds to the substrate concentration for half the enzyme molecules to bind to the substrate, therefore causing the reaction to proceed at half its maximum rate. K_m changes depending upon substrate. Low K_m value means binding affinity of the substrate to the enzyme is strong. For example, K_m for hexokinase enzyme; for glucose $K_\mathrm{m} = 0.05$ mM, and for fructose $K_\mathrm{m} = 1.5$ mM, which means that the enzyme reaction rate reaches its maximum with less amount of glucose compared to the case of fructose.

k_2 is often preferably substituted with k_cat. k_cat known as turnover number means the maximum number of substrate molecules that the enzyme can turnover to product in a set time. The ratio $k_\mathrm{cat}/K_\mathrm{m}$, known as catalytic efficiency, shows that the relative rate of reaction at low substrate concentration can be also interpreted as the utilization of different substrates for an enzyme.

4.1.3 Parameter Estimation

For enzymes to be used most efficiently, it is important to know the characteristics of enzymes. The kinetic parameters $V_{\max}$, K_{m}, and $k_{\mathrm{cat}}/K_{\mathrm{m}}$ should be therefore known. There are two approaches to this determination: the initial rates of reaction (differential method) or the reaction progress curve (integral method).

Equation 4.7 can be linearized in double-reciprocal form given as

$$\frac{1}{\upsilon} = \frac{1}{V\mathrm{max}} + \frac{K_{\mathrm{m}}}{V\mathrm{max}}\frac{1}{[S]}. \tag{4.8}$$

Plotting $1/\upsilon$ versus $1/[S]$ will give a linear line with a slope of $K_{\mathrm{m}}/V_{\max}$ and y-axis intercept of $1/V_{\max}$, as shown in Fig. 4.2, which is called as Lineweaver–Burk plot. Other plots, such as Eadie–Hostee plot and Haues–Woolf plot, can be used for Michaelis–Menten kinetics.

Alternatively, the time course of variation of $[S]$ in an enzyme reaction can be determined from

$$\upsilon = -\frac{\mathrm{d}[S]}{\mathrm{d}t} = \frac{V\mathrm{max}[S]}{K_{\mathrm{m}} + [S]} \tag{4.9}$$

By integration, using the boundary condition that at time zero substrate concentration is $[S_0]$, it yields

$$t = \frac{K_{\mathrm{m}}}{V\mathrm{max}}\ln\frac{[S_0]}{[S]} + \frac{1}{V\mathrm{max}}([S_0] - [S]). \tag{4.10}$$

Equation 4.10 is a form $t = a + bx_1 + cx_2$, and the constants a, b, and c can be obtained using least-squares method. From the constants, kinetic parameters can be also obtained. The parameter values of the equation can be used to predict the time for the substrate concentration reaches $[S]$ from $[S_0]$ and for the design and optimization of the reactor (Table 4.1).

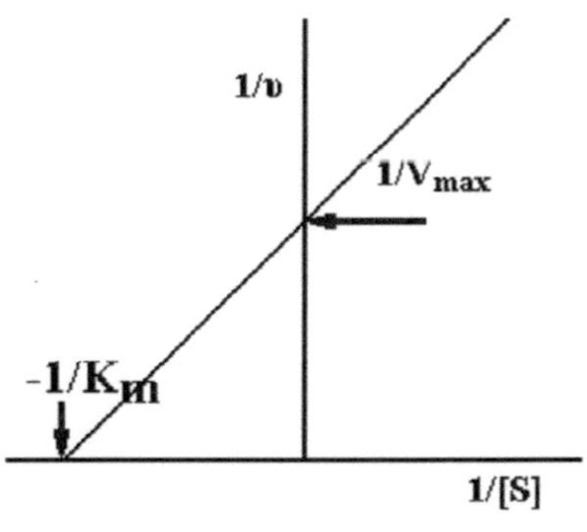

Fig. 4.2 Lineweaver–Burk plot

Table 4.1 Example of k_{cat} and K_m for enzyme–substrate pairs

Enzyme	Species	Substrate	k_{cat}	K_m
Alcohol dehydrogenase	*Aeropyrum pernix*	Ethanol	0.233	13.7
Alcohol dehydrogenase	*Aeropyrum pernix*	1-Butanol	0.411	0.596
Alcohol dehydrogenase	*Aeropyrum pernix*	1-Hexanol	0.368	0.147
Alcohol dehydrogenase	*Oenococcus oeni*	Ethanol	5.2	10.4
Triose-phosphate isomerase	*Gallus gallus*	D-glyceraldehyde 3-phosphate	3190	0.29
Triose-phosphate isomerase	*Gallus gallus*	Dihydroxyacetone phosphate	563	0.97
Triose-phosphate isomerase	*Saccharomyces cerevisiae*	D-glyceraldehyde 3-phosphate	8700	1.5
Triose-phosphate isomerase	*Saccharomyces cerevisiae*	Dihydroxyacetone phosphate	1725	2.3
Carboxylesterase	*Ferroplasma acidiphilum*	4-Nitrophenyl butyrate	55	0.32
Carboxylesterase	*Ferroplasma acidiphilum*	4-Nitrophenyl acetate	13	0.73
Carboxylesterase	*Sulfolobus solfataricus*	4-Nitrophenyl butyrate	4.7	0.93
Carboxylesterase	*Sulfolobus solfataricus*	4-Nitrophenyl caprylate	1.8	0.028

4.2 Other Enzyme Kinetics

4.2.1 Inhibition Kinetics

Inhibited Enzyme Reaction Kinetics. Enzyme inhibitors are compounds that reduce an enzyme reaction rate. The loss of enzyme activity may be reversible or irreversible. Reversible inhibitors dissociate more easily from the enzyme after binding to the enzyme through which the activity can be recovered. However, irreversible inhibitors (e.g., mercury and lead) can bind strongly to the amino acid backbone of the enzyme causing time-dependent loss of enzyme activity. There are three types of inhibitions—competitive, noncompetitive, and uncompetitive.

Competitive inhibition. *Inhibitors* compete with the substrate for the active site of the enzyme. The reaction involving competitive inhibitors can be modeled by the scheme

$$\begin{array}{l} E + S \underset{k_{-1}}{\overset{k_1}{\rightleftarrows}} ES \xrightarrow{k_2} E + P \\ +I \updownarrow K_I \\ \quad EI \end{array} \tag{4.11}$$

Assuming rapid equilibrium and with definition of

$$K_m = \frac{[E][S]}{[ES]}, \quad K_I = \frac{[E][I]}{[ES]}$$
$$[E_0] = [E] + [ES] + [EI], \quad \upsilon = k_2[ES]. \tag{4.12}$$

The following equation for the enzyme reaction rate can be developed:

$$\upsilon = \frac{V\mathrm{max}[S]}{K_m\left(1 + \frac{[I]}{K_\mathrm{I}}\right) + [S]}. \tag{4.13}$$

Equation 4.13 can be linearized in double-reciprocal form given as

$$\frac{1}{\upsilon} = \frac{1}{V\mathrm{max}} + \frac{K_\mathrm{m}\left(1 + \frac{[I\}}{K_\mathrm{I}}\right)}{V\mathrm{max}} \frac{1}{[S]} \tag{4.14}$$

Compared to the double reciprocal of the Michaelis–Menten and Fig. 4.2, the y-intercept remains the same but the slope changes by the factor of $\left(1 + \frac{[I\}}{K_\mathrm{I}}\right)$ (Fig. 4.3).

A similar effect is observed when different substrates coexist, as often found in industrial conversions. The reaction involved with two coexisting substrates each leading to a different product is shown by the following scheme:

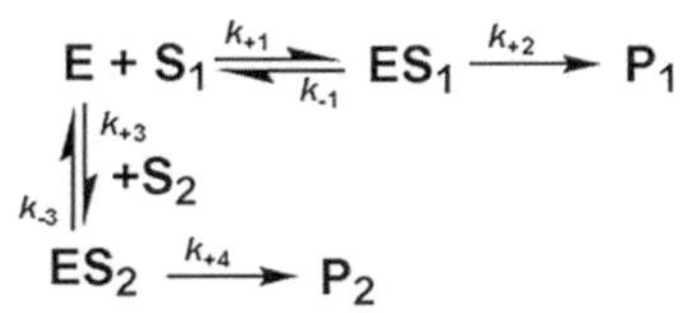

(4.15)

The substrates compete for the same catalytic site and thus they behave as competitive inhibitors of each other's reaction. The P_1 formation rate is given by

$$\upsilon_1 = \frac{V\mathrm{max}_1[S_1]}{K_{\mathrm{m}1}\left(1 + \frac{[S_2]}{K_{\mathrm{m}2}}\right) + [S_1]}, \tag{4.16}$$

and the P_2 formation rate is given by

$$\upsilon_2 = \frac{V\mathrm{max}_2[S_2]}{K_{\mathrm{m}2}\left(1 + \frac{[S_1]}{K_{\mathrm{m}1}}\right) + [S_2]}. \tag{4.17}$$

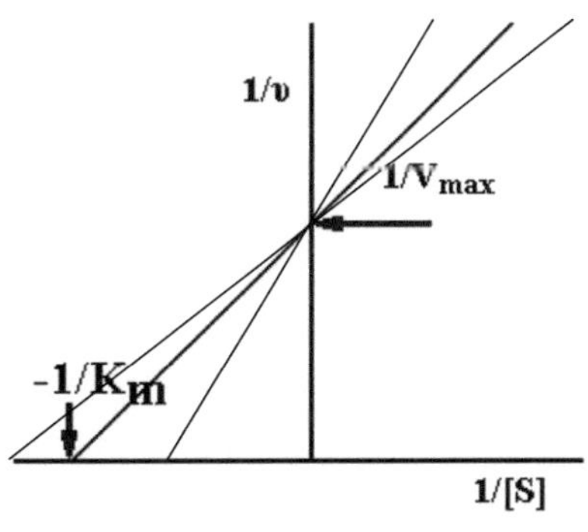

Fig. 4.3 Double-reciprocal plot for competitive inhibition

Noncompetitive inhibition. *Inhibitors* are not substrate analogs. These inhibitors attach on sites of other than the active site but decrease the enzyme's affinity to the substrate. The scheme for noncompetitive inhibitions is described as follows:

$$\begin{array}{ccccc} E+S & \overset{K'm}{\longleftrightarrow} & ES & \overset{k_2}{\longrightarrow} & E+P \\ +I\updownarrow K_I & & \updownarrow +I & & \\ EI & \underset{K'm}{\longleftrightarrow} & ESI & & \end{array} \tag{4.18}$$

With the definition of

$$K'_m = \frac{[E][S]}{[ES]} = \frac{[EI][S]}{[ESI]}, \quad K_I = \frac{[E][I]}{[\mathrm{EI}]} = \frac{[ES][I]}{[ESI]} \tag{4.19}$$

$$[E_0] = [E] + [ES] + [EI] + [ESI], \quad \upsilon = k_2[ES]. \tag{4.20}$$

The rate equation will be

$$\upsilon = \frac{V\mathrm{max}}{\left(1+\frac{[I]}{K_\mathrm{I}}\right)\left(1+\frac{K'_\mathrm{m}}{[S]}\right)}. \tag{4.21}$$

Uncompetitive inhibition. *Inhibitors* bind to the ES complex not to the enzyme itself. Uncompetitive inhibition can be described as follows:

$$\begin{array}{ccccc} E+S & \overset{K'm}{\longleftrightarrow} & ES & \overset{k_2}{\longrightarrow} & E+P \\ & & +I\updownarrow K_I & & \\ & & ESI & & \end{array} \tag{4.22}$$

With the definition of

$$K'_m = \frac{[E][S]}{[ES]}, \quad K_\mathrm{I} = \frac{[ES][I]}{[ESI]} \tag{4.23}$$

$$[E_0] = [E] + [ES] + [ESI], \quad \upsilon = k_2[ES]. \tag{4.24}$$

The rate equation will be

$$\begin{aligned} \upsilon &= \frac{\frac{V\mathrm{max}}{\left(1+\frac{[I]}{K_\mathrm{I}}\right)}[S]}{\frac{K'_\mathrm{m}}{\left(1+\frac{[I]}{K_\mathrm{I}}\right)}+[S]} \\ &= \frac{V\mathrm{max}[S]}{K'_\mathrm{m} + \left(1+\frac{[I]}{K_\mathrm{I}}\right)[S]}. \end{aligned} \tag{4.25}$$

Equation 4.25 can be linearized in double-reciprocal form given as

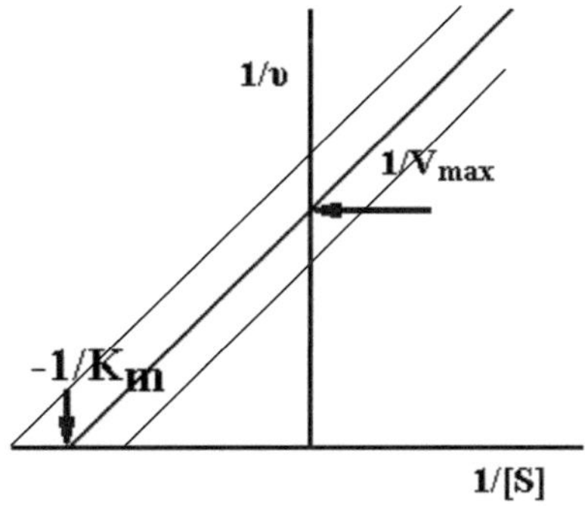

Fig. 4.4 Double-reciprocal plot for uncompetitive inhibition

$$\frac{1}{\upsilon} = \frac{1}{V\mathrm{max}}\left(1 + \frac{[I]}{K_I}\right) + \frac{K'_\mathrm{m}}{V\mathrm{max}}\frac{1}{[S]}. \tag{4.26}$$

Compared to the double-reciprocal plot of Michaelis–Menten equation and Fig. 4.2, the slope remains the same but the y-intercept changes depending upon $[I]$ by the factor of $\left(1 + \frac{[I]}{K_\mathrm{I}}\right)$. (Fig. 4.4).

Substrate inhibition kinetics. Another case of uncompetitive inhibition is substrate inhibition. This inhibition is usually observed at high concentrations of specific substrates. The substrate can bind to the enzyme–substrate complex which results in inactive complex and a decreased reaction rate.

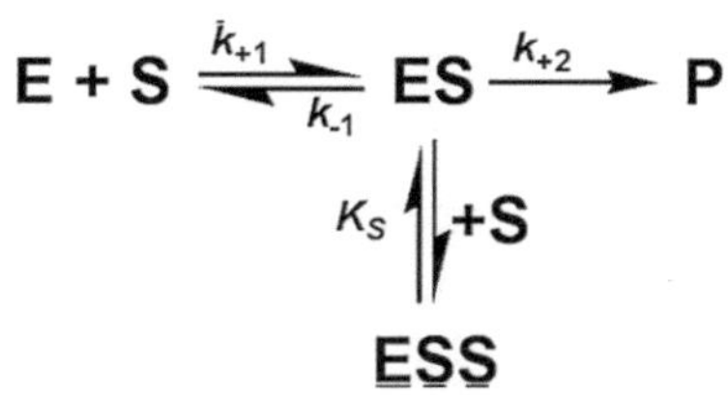

(4.27)

The rate equation when substrate inhibition occurs will be

$$\upsilon = \frac{V\mathrm{max}[S]}{K_\mathrm{m} + [S]\left(1 + \frac{[S]}{K_\mathrm{S}}\right)}. \tag{4.28}$$

$V_{\max}$ exists when $[S] = \sqrt{K_\mathrm{S}K_\mathrm{m}}$ (Fig. 4.5).

4.2.2 Other Kinetics

Reversible Reaction Kinetics. The reaction of glucose to fructose, catalyzed by glucose isomerase, follows *reversible enzymatic reaction.* This reversible reaction can be represented by the following scheme: enzyme–substrate complex

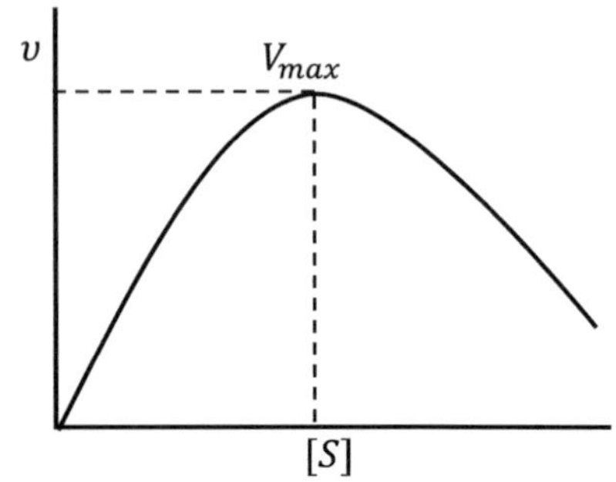

Fig. 4.5 Substrate inhibition kinetics curve

formation, conversion to enzyme–product complex and then desorption of the product:

$$E + S \underset{k_{-1}}{\overset{k_1}{\longleftrightarrow}} ES \underset{k_{-2}}{\overset{k_2}{\longleftrightarrow}} EP \underset{k_{-3}}{\overset{k_3}{\longleftrightarrow}} E + P. \tag{4.29}$$

The total enzyme concentration remains constant thus

$$[E_0] = [E] + [ES] + [EP]. \tag{4.30}$$

Following equations can be obtained for the intermediates *ES* and *EP* with the quasi steady-state assumption:

$$[ES] = \frac{(k_1k_{-2} + k_1k_3)[E_0][S] + k_{-2}k_{-3}[E_0][P]}{(k_{-1}k_{-2} + k_{-1}k_3 + k_2k_3) + (k_1k_2 + k_1k_{-2} + k_1k_3)[S] + (k_{-1}k_{-3} + k_2k_{-3}} \tag{4.31}$$

$$[EP] = \frac{k_1k_2[E_0][S] + (k_{-1}k_{-3} + k_2k_{-3})[E_0][P]}{(k_{-1}k_{-2} + k_{-1}k_3 + k_2k_3) + (k_1k_2 + k_1k_{-2} + k_1k_3)[S] + (k_{-1}k_{-3} + k_2k_{-3}} \tag{4.32}$$

The rate of reaction may be denoted by

$$\upsilon = \frac{\mathrm{d}[P]}{\mathrm{d}t} = k_2[ES] - k_{-2}[EP]. \tag{4.33}$$

Substituting from Eqs. 4.31 and 4.32 yields

$$\begin{aligned}\upsilon &= \frac{k_1k_2k_3[E_0][S] - k_{-2}k_{-3}[E_0][P]}{(k_{-1}k_{-2} + k_{-1}k_3 + k_2k_3) + (k_1k_2 + k_1k_{-2} + k_1k_3)[S] + (k_{-1}k_{-3} + k_2k_{-3}} \\ &= \frac{a(S) - b(P)}{c + d(S) + e(P)},\end{aligned} \tag{4.34}$$

where $a = k_1k_2k_3[E_0]$, $b = k_{-2}k_{-3}[E_0]$, $c = (k_{-1}k_{-2} + k_{-1}k_3 + k_2k_3)$, $d = (k_1k_2 + k_1k_{-2} + k_1k_3)$, $e = (k_1k_2 + k_1k_{-2} + k_1k_3)$.

Bisubstrate Reaction Kinetics. Many enzyme reactions involve two or more substrates. Examples are hydrolases (e.g., lipases) that utilize oil, alcohol, and water for esterification or transesterification reactions. Peptidases also require two substrates as well as lyases.

Ordered sequential Bi Bi mechanism. This mechanism is analyzed by considering an addition of two substrates (A and B) and the release of two products (P and Q):

$$A + B + E \underset{k_{-1}}{\overset{k_1}{\longleftrightarrow}} EA + B \underset{k_{-2}}{\overset{k_2}{\longleftrightarrow}} EAB \underset{k_{-3}}{\overset{k_3}{\longleftrightarrow}} EPQ \underset{k_{-4}}{\overset{k_4}{\longleftrightarrow}} P + EQ \underset{k_{-5}}{\overset{k_5}{\longleftrightarrow}} E + Q. \quad (4.35)$$

The EAB and EPQ complexes can be lumped together as EXY:

$$A + B + E \underset{k_{-1}}{\overset{k_1}{\longleftrightarrow}} EA + B \underset{k_{-2}}{\overset{k_2}{\longleftrightarrow}} \underbrace{\boxed{EAB \underset{k_{-3}}{\overset{k_3}{\longleftrightarrow}} EPQ}}_{EXY} \underset{k_{-4}}{\overset{k_4}{\longleftrightarrow}} P + EQ \underset{k_{-5}}{\overset{k_5}{\longleftrightarrow}} E + Q \quad (4.36)$$

With the quasi steady-state assumption, the following equation can be derived:

$$[E_0] = [E] + [EA] + [EXY] + [EQ] \quad (4.37)$$

$$\frac{\mathrm{d}[EA]}{\mathrm{d}t} = k_1[E][A] + k_{-2}[EXY] - k_{-1}[EA] - k_2[EA][B] = 0 \quad (4.38)$$

$$\frac{\mathrm{d}[EXY]}{\mathrm{d}t} = k_2[EA][B] + k_{-4}[EQ][P] - (k_{-2} + k_4)[EXY] = 0 \quad (4.39)$$

$$\frac{\mathrm{d}[EQ]}{\mathrm{d}t} = k_4[EXY] + k_{-5}[E][Q] - k_{-4}[EQ][P] - k_5[EQ] = 0. \quad (4.40)$$

Overall reaction rate can be obtained using above equations. These equations can be solved in several ways and be used.

Ping Pong Mechanism. In this mechanism, the first step of a cascade of reaction is carried out and the product is released before the substrate of the second step can bind. As shown in the scheme below, a free enzyme E and the first substrate, A, bind and undergo the first reaction. Product release results in a modified enzyme F which binds to the second substrate B for the second reaction. When the second product Q is released, the enzyme is recycled back to the E form. The first binary complexes EA and FP and the second binary complexes FB and EQ are lumped and represented as EX and FY, respectively.

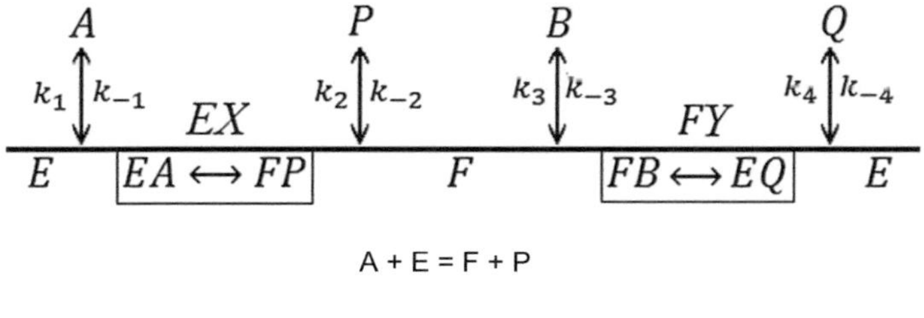

(4.41)

A + E = F + P

F + B = E + Q

Overall reaction is A + B = P + Q

Derivation gives the following equations for the following enzyme forms:

$$[E] = k_3k_4(k_{-1} + k_2)[B] + k_{-1}k_{-2}(k_{-3} + k_4)[P] \tag{4.42}$$

$$[EX] = k_1k_3k_4[A][B] + k_1k_{-2}(k_{-3} + k_4)[A][P] + k_{-2}k_{-3}k_{-4}[P][Q] \tag{4.43}$$

$$[F] = k_1k_2(k_{-3} + k_4)[A] + k_{-3}k_{-4}(k_{-1} + k_2)[Q] \tag{4.44}$$

$$[FY] = k_1k_2k_3[A][B] + k_3k_{-4}(k_{-1} + k_2)[B][Q] + k_{-1}k_{-2}k_{-4}[P][Q]. \tag{4.45}$$

Since $[P]$ and $[Q]$ are zero at the initial condition, this reaction can be expressed as

$$\upsilon = \frac{V\text{max}}{1 + \frac{K_a}{[A]} + \frac{K_b}{[B]}}, \tag{4.46}$$

where

$$V\text{max} = \frac{k_2k_4[E_0]}{k_2 + k_4} \tag{4.47}$$

$$K_a = \frac{k_{-4}(k_{-1} + k_2)}{k_1(k_2 + k_4)} \tag{4.48}$$

$$K_b = \frac{k_2(k_{-3} + k_4)}{k_3(k_2 + k_4)}. \tag{4.49}$$

Applications

The kinetic equations can be used to discriminate the kinetics or to find the reaction mechanism and for the design and optimization of the enzyme reactor system. Series enzymatic reactions are widely found in living cells as well as single-step reactions. The whole enzymatic reaction is regulated by feedback inhibition mode, where the product from last reaction step regulates the enzyme of the first reaction step. Also product inhibitions are widely found in many enzymatic reactions, which means that if product concentration is high enough, the enzyme reaction is then no more necessary in living cells. However, for industrial and economic applications, product inhibition should be minimized to obtain high product concentration. For this purpose many approaches have been suggested so far, such as engineering the enzymes to reduce the inhibition by changing the site of the enzyme where the product affects or binds, and in situ removal of the product or adopting efficient bioreactors. For refined practice, mathematical modeling of the reaction scheme is required.

4.2.3 pH Optimum of Enzymes

The reaction rate is affected by temperature, pH, and other environmental factors. Enzyme reaction rate generally increases as the reaction temperature increases

Table 4.2 Application of acidic and alkaline enzymes

Enzymes	Application
Alkaline protease	Detergent, dehairing, silk degumming, photographic gelatin hydrolysis to recover silver, feed, leather, baking, brewing, cosmetic, pharmaceuticals
Alkaline xylanase	Pulp bleaching, detergent, fuel alcohol production
Acidic xylanase	Food and feed
Acidic phytase	Feed
Acidic amylase	Glucose and fructose production
Alkaline amylase	Detergent
Alkaline lipase	Detergent, oleochemical
Acidic lipase	Food, flavor, acid bating of fur and wool, enzyme therapy
Alkaline cellulose	Pulp and paper industry
Acidic cellulose	Deinking (paper recycling)

until denaturation of the enzyme, where irreversible structural change in active site occurs. Also pH affects the enzyme reaction rate and pH optimum is usually observed, where highest reaction rate occurs.

Use of enzymes for industrial applications often requires harsh conditions of extreme pH. Some enzymes working in alkaline or acidic conditions have been utilized in a variety of applications (Table 4.2). However, most enzymes in nature have their maximum activity in around neutral pH conditions. Therefore, improvement of enzyme function by controlling the pH dependence of enzymatic catalysis is of importance for industrial application of enzymes.

The change of pH in the reaction solution shifts the equilibrium concentrations of the protonated and deprotonated states of the titratable residues. This modification changes the average charge of the residues. Electrostatic interactions are the primary factors upon which pH-related phenomena are dependent. Ionizable residues play key roles in enzymatic catalysis. The enzyme-mediated catalysis results from the ability of ionizable groups to function as nucleophilic, electrophilic, or general acid–base catalysts. The pH-dependent activity is closely related to the pK_a values of key ionizable groups within its active site (Neves-Petersen *et al.* 2001). It is important to define the factors that determine the pK_a values of the catalytic groups, upon which optimum pH of enzymes sets. Understanding of the factors is also helpful for the engineering of enzymes with tailored pH optima. Ionizable groups in proteins are necessary for catalysis and most biological energy transduction. During the catalytic cycle, the internal ionizable groups undergo various microenvironments, thus variable pK_a values and charged states (Rastogi and Girvin 1999). In highly polar microenvironments, most ionizable groups will be charged. In less polar microenvironments, neutral forms will be favored. In this case, the pK_a values of acidic groups will tend to be higher than usual (Dwyer *et al.* 2000; Karp *et al.* 2007), while basic groups have lower pK_a than the normal values (Fitch *et al.* 2002):

$$\mathrm{pH_{opt}} = \frac{\mathrm{p}K_1 + \mathrm{p}K_2}{2}, \tag{4.50}$$

where $\mathrm{p}K_1$ and $\mathrm{p}K_2$ are pK values of the main catalytic residues.

For obtaining various pH-optimum enzymes, screening from nature, screening using metagenomics, and engineering of enzymes are required. In the case of engineering enzymes for different pH optimums, understanding of pK value is essential.

Further Discussion

1. For reversible enzyme reaction, $S \leftrightarrow P$, how is equilibrium concentration determined? How can equilibrium concentration be changed?
2. Can we engineer the enzyme to increase k_{cat} value?
3. Can we engineer the enzyme to change K_{m} value for higher catalytic efficiency?
4. For alcohol dehydrogenase enzyme in Table 4.1, calculate catalytic efficiencies for different substrates and discuss the meaning of the result.
5. For inhibition kinetics (substrate, product), can we engineer the enzyme or enzyme reaction system to reduce the inhibition of the enzyme?

References

Dwyer JJ, Gittis AG, Karp DA, Lattman EE, Spencer DS, Stites WE and García-Moreno EB. High apparent dielectric constants in the interior of a protein reflect water penetration. *Biophysical Journal*, 2000, 79:1610–1620.

Fitch CA, Karp DA, Lee KK., Stites WE, Lattman EE and Bertrand García-Moreno E. Experimental pKa values of buried residues: Analysis with continuum methods and role of water penetration. *Biophysical Journal*, 2002, 82:3289–3304.

Karp DA, Gittis AG, Stahley MR, Fitch CA, Stites WE and García-Moreno EB. High apparent dielectric constant inside a protein reflects structural reorganization coupled to the ionization of an internal Asp. *Biophysical Journal*, 2007, 92:2041–2053.

Neves-Petersen MT, Petersen EI, Fojan P, Noronha M, Madsen RG and Petersen SB. Engineering the pH-optimum of a triglyceride lipase: from predictions based on electrostatic computations to experimental results. *Journal of Biotechnology*, 2001, 87:225–254.

Rastogi VK and Girvin ME. Structural changes linked to proton translocation by subunit c of the ATP synthase. *Nature*, 1999, 402(6759):263–268.

Chapter 5
Regeneration of Cofactors

Enzymes such as oxidoreductases and transferases are able to catalyze industrially useful reactions. However, these enzymes are often cofactor dependent. Cofactors are relatively low molecular weight compounds that are required for the enzymatic reactions. Examples of cofactors, as shown in Table 5.1, are several organic compounds, ATP, FAD, coenzyme A or the nicotinamide cofactors such as NAD(H) , and NADP(H) . Among cofactors, NAD(H) , and NADP(H) have been investigated extensively in recent years for applications to the synthesis of chemicals. Cofactors, since they act as stoichiometric agents in the enzyme reactions, react with substrates. Since these cofactors are often very expensive to be used as stoichiometric agents, efficient in situ regeneration of the reacted cofactors is required in industry (Figs. 5.1 and 5.2).

Methods for the regeneration of nicotinamide cofactors includes: enzymatic, chemical, photochemical, or electrochemical. Enzymatic methods can be either through one enzyme utilizing both the reduced and oxidized forms of a cofactor via coupling the desired product synthesis with cofactor regeneration reaction, as in Fig. 5.3a, or through two enzymes each working on the product synthesis and on the cofactor regeneration as in Fig. 5.3b.

Several requirements such as (1) practical and inexpensive process, (2) stable regeneration system, (3) easy separation of product, and (4) negligible byproduct formation, are to be considered to develop an efficient regeneration system.

5.1 Regeneration of Reduced Nicotinamide Cofactors

5.1.1 Enzymatic Method

Several enzymes are known to catalyze the regeneration of reduced nicotinamide cofactors. The important ones are given in Table 5.2.

Y.J. Yoo *et al.*, *Fundamentals of Enzyme Engineering*,
DOI 10.1007/978-94-024-1026-6_5

Table 5.1 Examples of cofactors and their regeneration methods (Zhao and van der Donk 2003)

Cofactor	Reaction type	Representative regeneration method
NAD	Removal of hydrogen	Glutamate dehydrogenase with α-ketoglutarate
NADH	Addition of hydrogen	Formate dehydrogenase with formate
NADP	Removal of hydrogen	Glutamate dehydrogenase with α-ketoglutarate
NADPH	Addition of hydrogen	Glucose dehydrogenase with glucose
ATP	Phosphoryl transfer	Acetate kinase with acetyl phosphate
Sugar nucleotides	Glycosyl transfer	Bacterial coupling
CoA	Acyl transfer	Phosphotransacetylase with acyl phosphate
PAPS	Sulfuryl transfer	Aryl sulfotransferase IV with *p*-nitrophenyl sulfate
S-Adenosyl methionine	Methyl transfer	No demonstrated method
Flavins	Oxygenation	Self-regeneration
Pyridoxal phosphate	Transamination	Self-regeneration
Biotin	Carboxylation	Self-regeneration
Metal porphyrin complexes	Peroxidation, oxygenation	Self-regeneration

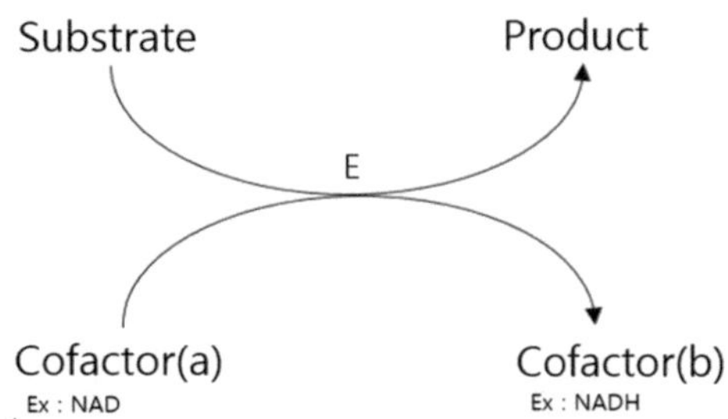

Fig. 5.1 Cofactor-mediated enzymatic reaction

Formate Dehydrogenase. Formate dehydrogenase (FDH) oxidizes formate to carbon dioxide and reduces NAD to NADH simultaneously. Carbon dioxide formed as a byproduct is chemically inert and can be emitted directly to the environment. Formate is cheap and commercially available as well as being innocuous towards most enzymes. Furthermore, the irreversibility of the reaction offers a major advantage for this enzyme for commercial uses (Fig. 5.4, Table 5.3).

Glucose dehydrogenases. Yun *et al.* (2003) investigated the synthesis of enantiomerically pure (*R*)-1-phenylethanol and (*R*)-α-methylbenzylamine from racemic α-methylbenzylamine using glucose dehydrogenase (GDH) for NADH regeneration. ω-Transaminase from *Vibrio fluvialis* SH1 and alcohol dehydrogenase (ADH) from *Lactobacillus kefir* were used for conversion in combination with

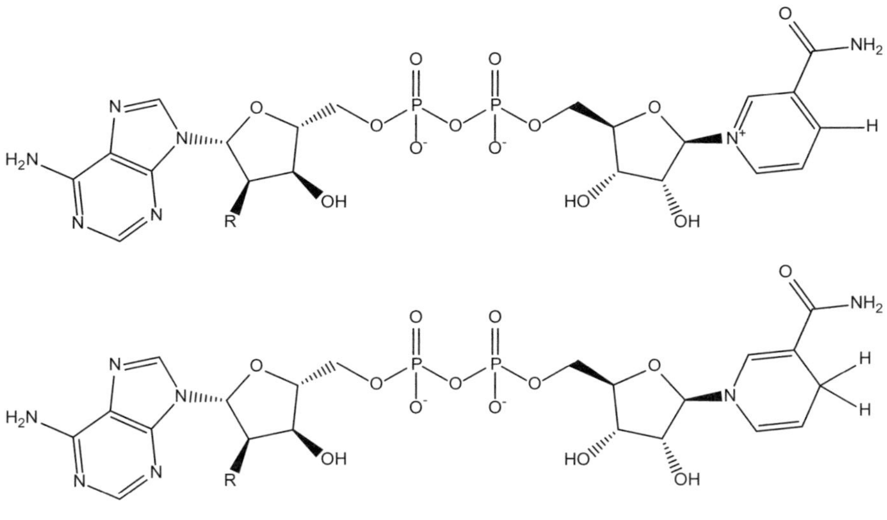

Fig. 5.2 Chemical structures of nicotinamide cofactors, NAD (*above*) and NADH (*below*), with an –OH group at R (NADP and NADPH have PO_4^{2-} instead of the –OH group at R)

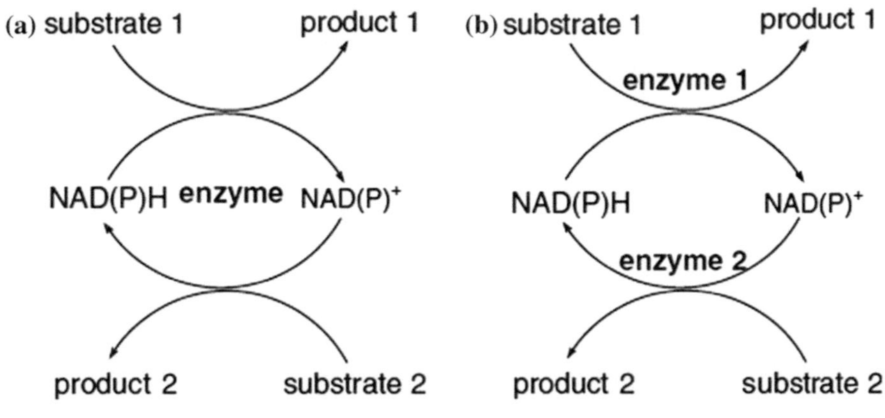

Fig. 5.3 Two enzymatic methods for the regeneration of nicotinamide cofactors

GDH from *Bacillus subtilis* (Fig. 5.5). The products of this reaction must be separated since gluconic acid from glucose is also of importance.

Alcohol Dehydrogenase. Alcohol dehydrogenase (ADH) can be used to catalyze reactions for the production of chiral compounds or for the regeneration of the coenzyme due to the reversibility of this reaction. ADH from *Thermoanaerobacter brockii* (TBADH) is often used for NADPH regeneration, by simultaneous oxidation of 2-propanol to acetone.

Hydrogenase. Hydrogen dehydrogenase can be used to reduce electron acceptors by hydrogen, and NAD as an electron acceptor can be reduced by the

Table 5.2 Enzymes used for the regeneration of reduced nicotinamide coenzymes (Weckbecker *et al.* 2010)

Enzyme EC number	Organism	Activity (U mg^{-1})
Formate dehydrogenase EC 1.2.1.2	*Candida boidinii*	4–6
	Candida methylica	4–6
	Candida methanolica	4–6
	Pseudomonas sp.	4–6
	Thiobacillus sp.	7.6
	Mycobacterium vaccae	
Glucose-6-phosphate dehydrogenase EC 1.1.1.49	*Leuconostoc mesenteroides* *Bacillus stearothermophilus*	290
Glucose dehydrogenase EC 1.1.1.47	*Thermoplasma acidophilum*	
	Bacillus megaterium	550
	Bacillus subtilis	375
Alcohol dehydrogenase (NAD^+) EC 1.1.1.1	*Saccharomyces cerevisiae*	
	Horse liver	
Alcohol dehydrogenase ($NADP^+$) EC 1.1.1.2	*Saccharomyces cerevisiae* *Thermoanaerobacter brockii* *Lactobacillus brevis* *Lactobacillus kefir*	
Hydrogenase EC 1.12.1.2	*Alcaligenes eutrophus*	54
	Hydrogenomonas H 16	
Hydrogenases EC 1.12.7.2	*Pyrococcus furiosus*	360
Phosphite dehydrogenase EC 1.20.1.1	*Pseudomonas stutzeri*	16

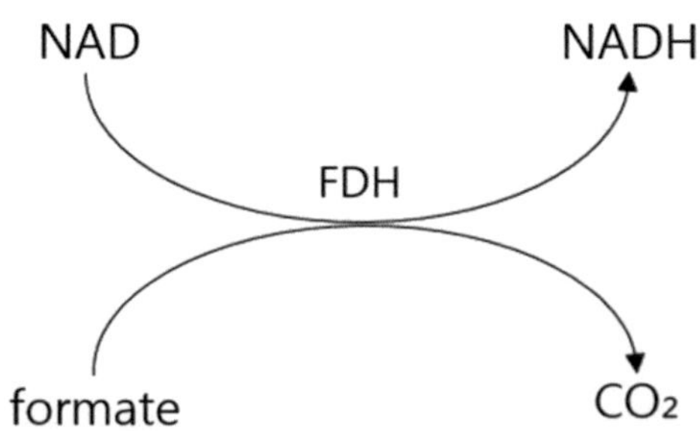

Fig. 5.4 Formate dehydrogenase (FDH) mediated oxidation of formate

hydrogen. Hydrogenases are considered in the regeneration processes since it can be easily isolated from microbial sources, the substrate hydrogen is cheap and the product separation is not affected by the regenerating system.

$$NADP + H_2 \underset{\text{hydrogenase}}{\longrightarrow} NADPH + H$$

5.1.2 Electrochemical Method

Electrochemical method employing enzymes and electrode are also being investigated to have various tools for cofactor regeneration. One example is shown in Fig. 5.6. The choice on what technology is adopted and used depends upon economics, easiness of operation, etc.

5.1.3 Application to Industrial Processes

Next step is to build a reactor system for cofactor regeneration. If the product is soluble in organic solvent, add organic solvent after the reaction. Since enzymes and NAD are water soluble, separation of the solvent phase stream which contains the product from the aqueous stream makes the recycle of the regeneration system.

Table 5.3 Enzymes for the regeneration of oxidized nicotinamide coenzymes (Weckbecker *et al.* 2010)

Enzyme EC number	Organism	Activity (U mg^{-1})
Glutamate dehydrogenase (NAD^+) EC 1.4.1.2	*Neurospora crassa*	
Glutamate dehydrogenase ($NADP^+$) EC 1.4.1.4	*Escherichia coli*	130 (reductive amination of ketoglutarate)
Glutamate dehydrogenase; dual specificity EC 1.4.1.3	*Bos Taurus*	167 (oxidative deamination of L-glu)
L-Lactate dehydrogenase EC 1.1.1.27	*Lactobacillus casei* *Lactobacillus lactis* *Lactobacillus curvatus* *Streptococcus epiderminis*	2290 (pyruvate reduction) 2030 (pyruvate reduction)
NADH oxidase (H_2O-forming) EC 1.6.3.1	*Lactobacillus brevis* *Lactobacillus sanfranci-scensis*	350
NADH oxidase (H_2O_2-forming) EC 1.5.99.3	*Escherichia coli*	5.2 (NADH oxidation)

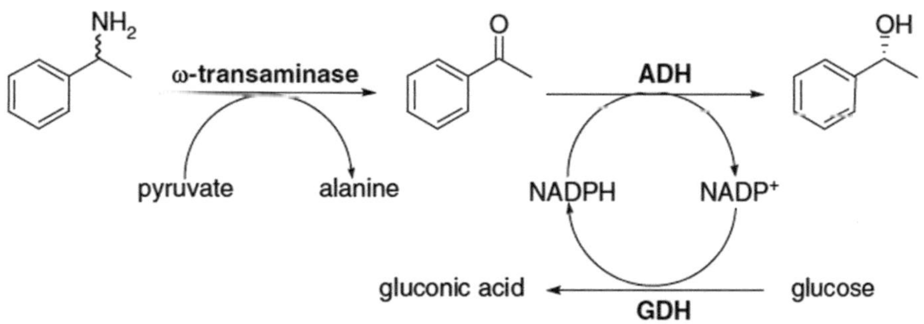

Fig. 5.5 Synthesis of (*R*)-phenylethanol and (*R*)-α-methylbenzylamine

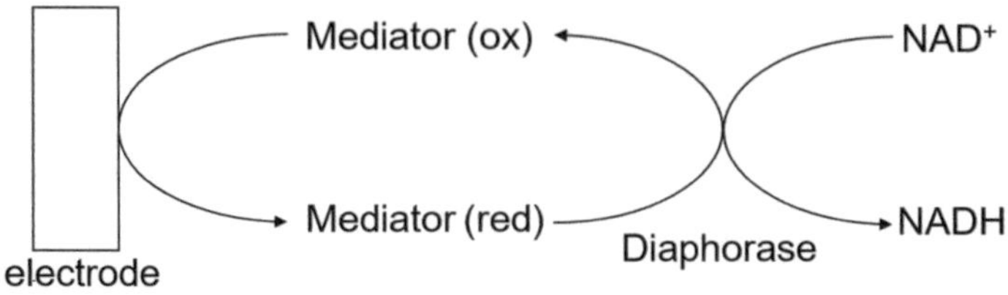

Fig. 5.6 Electrochemical regeneration of NADH

Membrane reactor system can be also used. In this case, cofactor can be remained inside of the membrane, once NAD can be combined with high molecular weight material such as PEG, where enzymes and PEG-NAD can be located inside of the membrane and substrate and product can be penetrated into and out of the membrane.

Carrea *et al.* (1988) coupled a NAD^+ regeneration system using lactate dehydrogenase/pyruvate to a hydroxysteroid dehydrogenase-mediated oxidation reaction. The glutamic dehydrogenase/2-oxoglutarate system and the alcohol dehydrogenase/acetaldehyde system were also used. These systems have a quite favorable equilibrium for the regeneration of NAD^+.

Baeyer–Villiger oxidations of prochiral ketones produce chiral lactones to be used for the synthesis of natural products. Rissom *et al.* (1997) synthesized ε-lactones using a cyclohexanone mono-oxygenase (CHMO) with NADPH regeneration system of an engineered formate dehydrogenase. 5-Methyloxepane-2-one was produced with a purity of >99% using a repetitive batch mode (see Fig. 5.7).

The poor solubility of some substrates in aqueous media led to the application of enzyme membrane reactors. Introduction of an emulsion membrane reactor can increase the substrate solubility while decreasing product inhibition via a coupled product separation step. Liese *et al.* (1998) showed that hydrophilic ultrafiltration membrane can separate an emulsion of 2-octanone and aqueous buffer solution, and the substrate dissolved in the aqueous phase is input into the emulsion

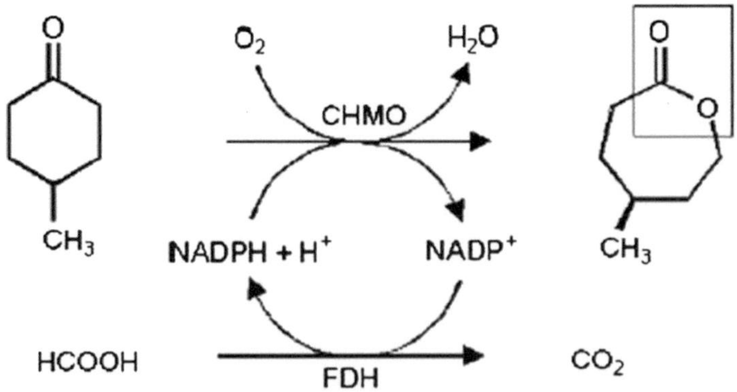

Fig. 5.7 Process for mono-oxygenase catalyzed Baeyer–Villiger reaction. The reaction can be performed with repetitive batch mode—substrate filling, reaction, and separation of the product

membrane reactor. A carbonyl reductase reduces 2-octanone to (*S*)-2-octanol enantioselectively and the product is extracted via the organic phase. NADH is regenerated simultaneously in the emulsion membrane reactor via formate dehydrogenase. (*S*)-2-octanol was produced with an enantiomeric excess of >99% during four-month operation of the reactor.

5.2 Regeneration of Oxidized Nicotinamide Cofactors

5.2.1 Enzymatic Method

To date, many enzymatic methods have been developed for the regeneration of oxidized nicotinamide cofactors (Fig. 5.8).

Glutamate dehydrogenase (GluDH) was used for $NADP^+$ regeneration in the synthesis of 12-ketochenodeoxycholic acid via 12α-hydroxysteriod dehydrogenase from *Clostridium* as in Fig. 5.9. L-Lactate dehydrogenase (L-LDH) can oxidize NADH to NAD by reducing pyruvate to (*S*)-lactate. This enzyme is very specific for pyruvate and short-chain 2-keto acids as substrates.

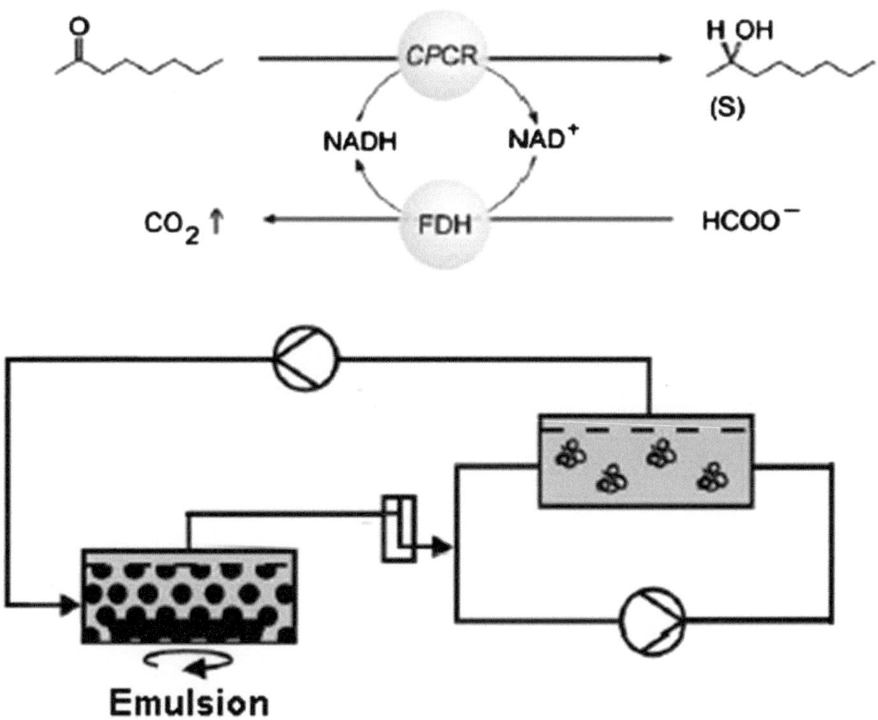

Fig. 5.8 Production of (*S*)-2-octanol with cofactor regeneration in emulsion enzyme membrane reactor (Liese *et al.* 1997)

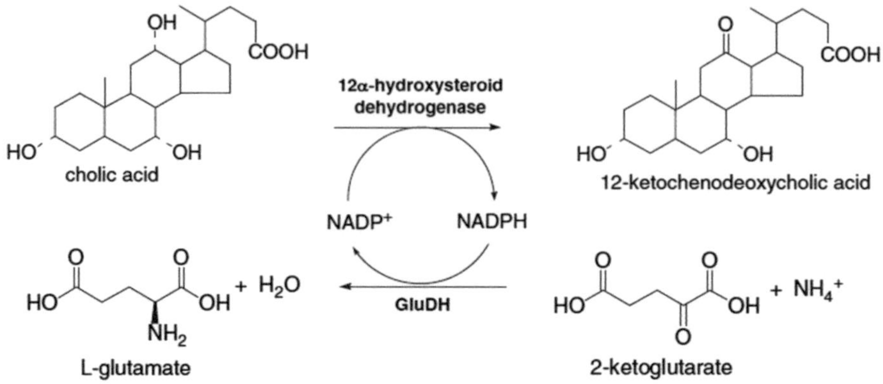

Fig. 5.9 Production of 12-ketochenodeoxycholic acid from cholic acid

5.2.2 Electrochemical Method

Electrochemical cofactor regeneration uses cheap electricity instead of enzyme, making it simple and cost effective even though an electrochemical apparatus is required. An easy recovery of the desired product is possible because no co-substrate, and therefore no byproduct, is present. The method for electrochemical regeneration of nicotinamide cofactors can be divided into two concepts: (a) direct regeneration and (b) mediated regeneration of cofactors (Fig. 5.10).

Fig. 5.10 Scheme of **a** direct and **b** and **c** mediated electrochemical regeneration of NAD^+

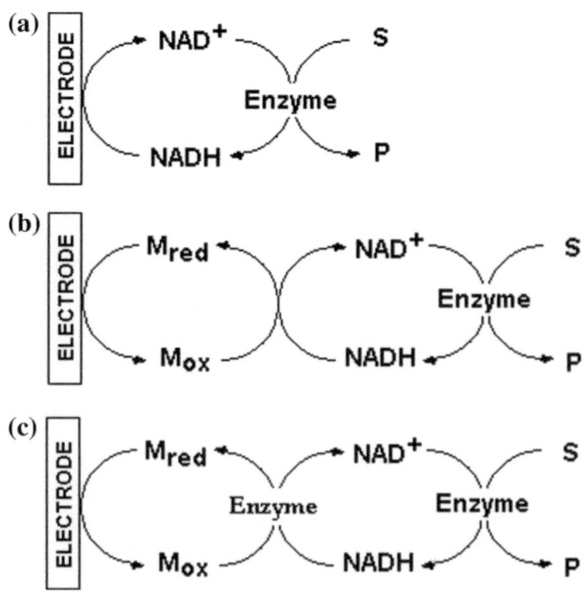

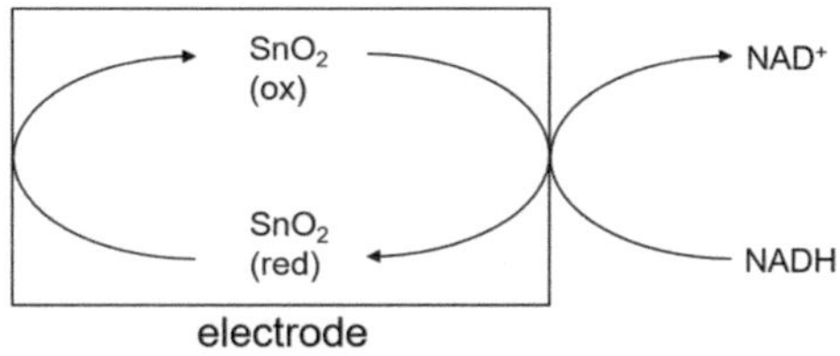

Fig. 5.11 Schematic diagram of cofactor NAD regeneration using tin oxide electrode

Direct electrochemical oxidation of NAD(P)H to $NAD(P)^+$ on the electrode surface can be successfully performed. However, relatively high oxidation potential is required which can lead to undesirable side reactions such as electrode passivation. The anodic potential depends on the electrode material, conditioning, and pretreatment methods used (Fig. 5.11).

In order to reduce the high potential in case of direct electrochemical regeneration, carbon electrodes are used because of their large surface. For the direct electrochemical oxidation of NAD(P)H, anode potentials should be larger than +900 mV compared to the normal hydrogen electrode (NHE). On the other hand, the introduction of an artificial electron transferring agent called mediator leads to an increase on electrons between NADH and the anode. Electron transfer occurs through the mediator under lowered overpotential than direct regeneration, thus diminish the occurrence of unwanted side reactions. However, the mediator might cause as an inhibitor to the enzyme and very expensive.

Recent advances. Direct electrochemical regeneration of a flavin-dependent monooxygenase enzyme was performed for opticaly pure epoxides synthesis (Hollman *et al.* 2005). New electrode material employing tin oxide electrode for the NADH oxidation under low overpotential without a mediator was suggested (Kim and Yoo 2009). For the oxidation of 2-propanol to acetone, $NADP^+$-dependent alcohol dehydrogenase was used. The reduced cofactor by the enzyme reaction is electrochemically reoxidized on the anodic tin oxide electrode. Conversion of 91% was obtained for 51 h reaction time. This electrochemical regeneration system was also applied for the kinetic resolution of (rac)-2-pentanol which is coupled to thermophilic alcohol dehydrogenase. The system obtains 50% conversion and an enantiomeric excess for (*R*)-2-pentanol of >99% after 9 h. TADH (thermophilic alcohol dehydrogenase) can convert (rac)2-pentanol to (*R*)2-pentanol and 2-pentanon using tin oxide electrode system where the mediator was not required.

5.2.3 Photochemical Method

Photochemical methods have been investigated to solve the problems in direct electrochemical oxidation of NAD(P)H. Photosensitizers that absorb visible light to convert the energy for electron transfer processes, such as organometallic complexes Ru(II)-*tris*-bipyridine $(Ru(bpy)_3)^{2+}$, metalloporphyrins

Zn-*meso*-tetramethyl-pyridinium porphyrin (Zn-TMPyP^{4+}), and organic dye compounds (*i.e.,* acridine or flavin dyes), and semiconductor materials such as TiO_2, CdS, or Fe_2O_3 in the form of powders or colloids are used to mediate electron transfer reactions (Willner and Mandler 1989).

Further Discussion

1. Since formate dehydrogenase (FAD) is very important for cofactor regeneration, much efforts have been made for the improvement of FAD. Discuss what attempts have been made to improve the properties of FAD?
2. How to regenerate ATP?
3. ATP and NAD(P) are very important and sometimes bottleneck for enzyme reaction. How ATP or NAD(P) pool are synthesized in living cells and can be increased?

References

Carrea G, Riva S, Bovara R and Pasta P. Enzymatic oxidoreduction of steroids in two-phase systems: effects of organic solvents on enzyme kinetics and evaluation of the performance of different reactors. *Enzyme and Microbial Technology*, 1988, 10:333–340.

Hollmann F, Hofstetter K, Habicher T, hauer B, Schmid A. Direct electrochemical regeneration of monooxygenase subunits for biocatalytic asymmetric epoxidation. *JACS*. 2005, 127:6540–6541.

Kim YH and Yoo YJ. Regeneration of the nicotinamide cofactor using a mediator-free electrochemical method with a tin oxide electrode. *Enzyme and Microbial Technology*, 2009, 44:129–134.

Liese A, Zelinski T, Kula MR, Kierkels H, Karutz M, Kragl and Wandrey CA. Novel reactor concept for the enzymatic reduction of poorly soluble ketones. *Journal of Molecular Catalysis B: Enzymatic*, 1998, 4:91–99.

Rissom S, Schwarz-Linek U, Vogel M, Tishkov VI and Kragl U. Synthesis of chiral E-lactones in a two-enzyme system of cyclohexanone mono-oxygenase and formate dehydrogenase with integrated bubble-free aeration. *Tetrahedron: Asymmetry*, 1997, 8:2523–2526.

Weckbecker A, Gro¨ger H and Hummel W. Regeneration of nicotinamide coenzymes: principles and applications for the synthesis of chiral compounds. *Advances in Biochemical Engineering/ Biotechnology*, 2010, 120:195–242.

Willner I and Mandler D. Enzyme-catalysed biotransformations through photochemical regeneration of nicotinamide cofactors. *Enzyme and Microbial Technology*, 1989, 11:467–483.

Yun H, Yang YH, Cho BK, Hwang BY and Kim BG. Simultaneous synthesis of enantiomerically pure (*R*)-1-phenylethanol and (*R*)-α-methylbenzylamine from racemic α-methylbenzylamine using ω-transaminase/alcohol dehydrogenase/glucose dehydrogenase coupling reaction. *Biotechnology Letters*, 2003, 25:809–814.

Zhao H and van der Donk WA. Regeneration of cofactors for use in biocatalysis. *Current Opinion in Biotechnology*, 2003, 14:583–589.

Chapter 6
Immobilized Enzyme

One of the factors that hinders the use of enzyme for industrial processes is their cost. Since enzymes are catalyst, they are not consumed during reactions. If enzymes are used as free form, they should be recovered after the reaction for reuses. The remaining enzymes might also contaminate the product if it is not separated during purification steps. Separating the enzyme from the product during or after the reaction using an aqueous/nonaqueous biphase system can be a solution. One phase contains the enzyme and the other phase contains the product. An example of two-phase system is immobilized enzymes: by fixing enzymes on or within support materials.

The important benefits derived from immobilization of enzymes are (1) ease of separation of the product from reaction medium and (2) reusability of the enzyme. Ease of separation of the enzyme from the product stream enables enzyme's industrial applications more efficient. Contamination of the product from the enzyme reaction can be prevented using immobilized enzyme which helps to reduce the downstream processing cost. Reusing the enzyme can provide cost advantages which are often an essential part for economically feasible commercial enzyme catalyzed processes (Tischer and Wedekind 1999). The comparison between immobilized enzymes and free enzymes is given in Table 6.1.

6.1 Methods of Enzyme Immobilization

6.1.1 Immobilization Methods

There are several methods for enzyme immobilization: physical methods and chemical methods. As shown in Fig. 6.1, physical methods include adsorption and entrapment; chemical methods include cross-linking and covalent binding

Y.J. Yoo *et al.*, *Fundamentals of Enzyme Engineering*,
DOI 10.1007/978-94-024-1026-6_6

Table 6.1 Comparison of immobilized enzymes with free enzymes

Free enzyme	Immobilized enzyme
Difficult to separate and reuse	No need to separate from the product stream
Enzymes are generally expensive	Can be reused several times, thus the cost for enzyme can be reduced
Activity is influenced by temperature, pH, and other operating conditions	Properties can be changed: activity, temperature profile, pH optimum

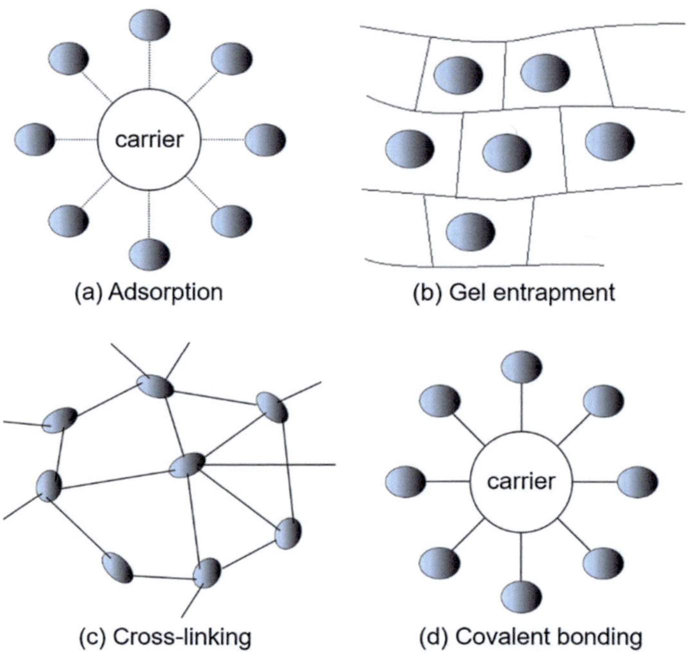

Fig. 6.1 Methods of enzyme immobilization

(Sheldon 2007). Regardless of the method of immobilization, the materials to be used for the enzyme immobilization should be insoluble in the reaction medium.

Immobilization of enzyme should offer good physical properties that are relevant to industrial applications although it requires the support material to facilitate the diffusion of the substrate and product. For example, a nanostructured biocatalyst support matrix composed of an enzyme-entrapped hydrophilic phase and a hydrophobic domain for substrate and product diffusion was introduced to increase the turnover number up to 230-fold for horseradish peroxidase-catalyzed coupling reaction in heptanes (Bruns and Tiller 2005).

Adsorption. This method involves the physical attachment of the enzyme onto the backbone or support material such as DEAE-cellulose, activated carbon, in this case the forces are weak, such as van der Waals, hydrophobic interactions or dispersion forces. Adsorption is still widely employed compared to other methods because of many advantages: (1) mild preparation conditions, (2) low cost, (3) no chemical reactions, and (4) high activity. However, since the force for the adsorption is usually weak, enzymes can be easily leached out and thus contaminate the product. In many cases, ion-exchange resin is widely used in industry to immobilized enzymes, where the interaction between the matrix and the enzyme is much stronger than adsorption.

Entrapment. Enzymes are captured within the cavities of a matrix or microcapsule of polymer. Synthetic polymers such as polyacrylamide and polyvinyl alcohol, natural polymers such as Ca-alginate and carrageenan can be used. Recently, sol–gel polymerization is used for this purpose where inorganic materials can be also employed. This method usually reduces chances of leaching out of enzymes and improves stabilization, but results in mass transfer limitations of the substrate and the product.

Cross-Linking. This method can provide a three-dimensional network of enzymes by coupling reagent. The enzymes become stable because of the strong interaction between the enzymes and the carrier, but the cross-linking reagents and conditions used may damage the enzymes.

Covalent Binding. The covalent attachment to a matrix, which is widely used, has the advantage of strong interactions between the enzyme and the carrier material which makes the enzyme very stable. $-NH_2$ group in lysine or arginine, –COOH group in aspartic acid or glutamic acid, –OH group in serine or threonine, and –SH group in cysteine can be used for cross-linking with the cross-linking reagents. Eupergit, commercial name, is known as one of the best immobilization matrices for high loading of enzymes on the matrix material, which is important to reduce the reactor size and to increase the reaction efficiency.

6.1.2 Recent Advances

In addition to these traditional methods, many other methods have been recently developed.

- Encapsulation of enzymes in various forms of semipermeable membranes such as nylon or cellulose nitrate was introduced.
- Display of enzymes on the surface of microorganisms such as yeast was introduced. This surface display method has also many advantageous since enzyme reaction and fermentation by microorganisms can be performed at the same time in the same reactor system.
- To prevent steric hinderance, site-specific immobilization was introduced (Hernandez and Fernandez-Lafuente 2011), where active sites of the enzyme are open to the substrates.
- Employing nanoparticles for enzyme immobilization matrix provides many advantages in properties and applications. Nowadays, nanotechnology has been

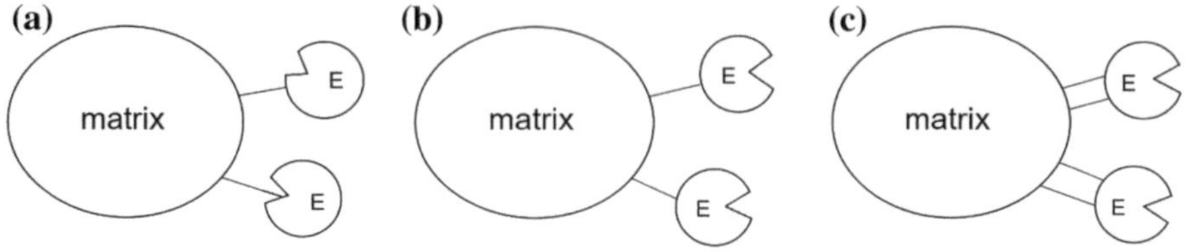

Fig. 6.2 Schematic diagram of site-specific immobilization **a** not desirable **b**, **c** desirable

Table 6.2 Comparison of different immobilization methods (Zhang *et al.* 2012)

Characteristics	Adsorption	Entrapment	Cross-linking	Covalent bonding
Immobilization	Easy	Difficult	Difficult	Difficult
Interaction	Weak	Strong	Strong	Strong
Recovery of activity	Low	High	Moderate	Low
Regeneration of immobilized enzyme	Possible	Impossible	Impossible	Impossible
Immobilization cost	Low	Low	Moderate	High

developed and is widely applied in many fields including enzyme immobilization (Min and Yoo 2014). For example, enzymes were immobilized as single enzyme nanoparticles form, which gave many advantages in immobilized enzyme applications including high stability (Kim and Grate 2003) (Fig. 6.2).

Following are the checkpoints for enzyme immobilization. (1) enzyme should maintain high activity and stability, (2) carrier material for the immobilization should have strong mechanical strength, (3) immobilized enzyme should have no or little steric hindrance of the enzyme, and (4) the cost for enzyme immobilization should be low. Table 6.2 gives the comparison of different immobilization methods.

6.2 Characteristics of Enzyme Immobilization

Immobilization of enzymes affects microenvironment of the enzyme and thus shows various characteristics, such as pH optimum, selectivity, stability, etc.

Enzyme Activity. Immobilized enzymes show in some cases higher activity than the free enzyme. This increase of the activity depends on many factors such as microenvironment, enzyme conformational change and orientation in the supporting matrix, and diffusion effects on substrates and products. For example, some lipases can take both the closed (inactive) form and the open (active) form. In the open form, the lid covering the entrance to the active site is shifted to expose the active site to the reaction medium. The immobilization support can be designed to control the conformation of lipase to its active form.

Thermal Stability. The stability of the immobilized enzyme depends on the immobilization condition, the amount and strength of the interaction with the support, binding position, flexibility, structure of the support and microenvironment. Encapsulation provides the highest possibility to lengthen enzymes' half-life, although mass transfer limitations can be a problem. Multipoint attachment of enzyme to a support decreases its structural flexibility and provides rigidity of the enzyme which is good for stability. Often times, the increase in enzyme stability through immobilization results in the decrease of the activity.

Solvent Stability. Organic solvents are detrimental to enzymes. Once enzymes make aggregates in organic solvents, substrate becomes inaccessible to the enzyme. However, immobilization of enzyme into mesoporous materials can improve their activity in organic solvents, if properly designed and selected, since the interaction between solvent and the enzyme can be changed to positive direction.

Selectivity. Enzyme-mediated asymmetric synthesis is in demand nowadays due to the increasing need of optically pure intermediates for pharmaceutical use. For example, a nonselective enzyme such as chloroperoxidase is transformed after immobilization into a stereoselective enzyme. The *S*-selective *Candida rugosa* lipase has also been converted to *R*-selective by covalent immobilization (Palomo *et al.* 2002). Immobilization may affect the enzyme structure, particularly in the active site to change the substrate selectivity. However, the theoretical and structural understanding on the improvement of enzyme selectivity by immobilization is not yet well known and should be investigated further.

6.2.1 *Mass Transfer Issues*

Despite of the mentioned advantages of immobilized enzymes, there are inherent limitations such as enzyme leakage into the reaction medium, diffusional resistance, decreased enzyme activity, loss enzyme activity during the immobilization and lack of appropriate control method for microenvironment. Reduced enzyme activity and loss of enzyme activity are caused mainly by unfavorable microenvironment surrounding enzymes. More favorable microenvironment can be made by optimizing matrix size and using different chemical compositions. Diffusion barrier problem is usually less severe in surface immobilized enzyme rather than encapsulated enzyme.

Whether diffusion barrier has an adverse effect on the reaction rate depends on the relative ratio between reaction rate and diffusion rate, which is represented by the Damkohler number (Da).

$$\mathrm{Da} = \frac{\text{maximum reation rate}}{\text{maximum diffusion rate}} = \frac{V_{\max}}{k_{\mathrm{L}}[S_{\mathrm{b}}]}, \tag{6.1}$$

where $[S_b]$ is substrate concentration in bulk solution and k_L is the mass transfer coefficient.

If Da $\gg$ 1, maximum reaction rate is much bigger than maximum diffusion rate, which implies that diffusion step is rate limiting. On the while, if Da $\ll$ 1, reaction step is rate limiting.

External mass transfer and internal mass transfer through the pores of immobilization support can affect or even dominate the observed reaction rate if Da number is much bigger. External mass transfer limitations take place if the diffusive transport rate of substrate or product through the stagnating layer is seriously lower. Internal mass transfer limitation in porous matrix of supports implies that transport of substrate or product from the surface of carrier to the active site of enzyme immobilized in the pore of support is the slowest step among overall mass transfer steps.

If external diffusion step is the rate determining one, remarkably decreased enzyme activity will be observed. The flux through the constant liquid film at the surface of immobilized enzyme can be expressed as the following equation:

$$\text{mass tranfer flux through liquid film} = (k_S/\delta) * ([S]_b - [S]_s), \tag{6.2}$$

where k_s is the mass transfer coefficient of the limiting substrate, δ is the thickness of the liquid film layer on the immobilized enzyme, $[S]_b$ and $[S]_s$ represent concentration of limiting substrate at bulk liquid and surface, respectively. The mass transfer coefficient k_s can be estimated by the relationship shown as follows:

$$\frac{k_S * d_P}{D} = 2 + c\left(\frac{d_P * v}{\nu}\right)^{0.5} * \left(\frac{\nu}{D}\right)^{1/3}, \tag{6.3}$$

where d_p is the diameter of immobilized enzyme particle, D is the diffusion coefficient of substrate, and ν is the dynamic viscosity.

External diffusion approximately shows the inversely proportional relationship with liquid film thickness (δ), which can be reduced significantly by vigorous stirring since compaction of film layer is caused. If the observed reaction rate is not improved upon enhanced mixing, external mass transfer may not be rate limiting step. After the verification of external diffusion effect as rate determining step, internal diffusion through pore of matrix must be concerned.

As shown in Fig. 6.3, effectiveness factor (ratio between reaction rate considering internal diffusion and intrinsic reaction rate neglecting diffusion effect) can be seriously influenced by the particle radius and enzyme contents. If the radius of immobilized enzyme particle gives significant impact on observed reaction rate, internal diffusion must be considered.

Case Study: Immobilization of *Candida antarctica* Lipase B, CALB (Yagonia *et al.*2014)

Silica-based carriers are commonly used in enzyme immobilization. The sol–gel technique was used to entrap the enzyme and to generate silica matrices using acid or base-catalyzed hydrolysis of hydrolysable silane compounds such as tetraethyl orthosilicate (TEOS), $Si(OC_2H_5)_4$. The chemical reaction in the sol–gel formation is presented in Fig. 6.4. The liquid precursor, $Si(OR)_4$, where R is CH_3 or C_2H_5 was hydrolyzed with water forming hydrated silica tetrahedral and alcohol

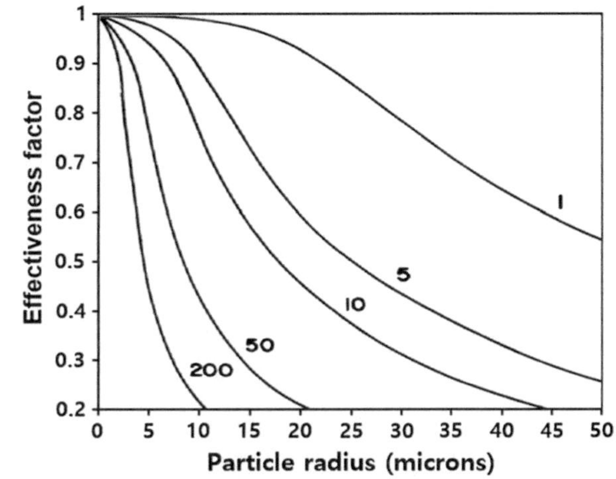

Fig. 6.3 Influence of radius on the effectiveness factor of a spherical porous immobilized enzyme particle (enzyme contents [mg cm^{-3}] are listed besides *curves*) (specific activity: 100 IU, $D_{eff} = 4 \times 10^{-8}$ m^2 s^{-1}, $[S]/K_M = 10$) (Regan 1974)

as by-product. The hydrated silica tetrahedral then reacted in a condensation reaction scheme to form ≡S–O–S≡ bonds. Linkage of additional ≡S–OH tetrahedral resulted in a SiO_2 network.

Reetz *et al.* (1995) efficiently immobilized lipases in hydrophobic silicon oxide sol–gel matrix by entrapment. They presented that enhanced enzyme activity correlates with the silicon oxide's higher hydrophobicity. Interfacial activation triggered by hydrophobic interactions led to a "lid-opened" state fixed in the matrix and thus constantly active form (Reetz 1997). Higher stability was also observed, due to multiple hydrogen bonding, hydrophobic interaction (van der Waals), and ionic interaction. However, enzymes are exposed to extreme environment during the entrapment where solvent is used. Accessibility of the substrates to the

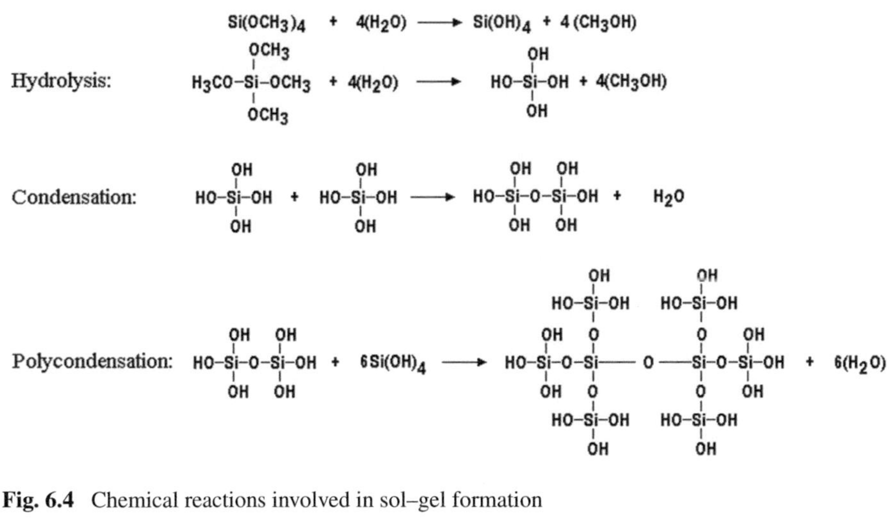

Fig. 6.4 Chemical reactions involved in sol–gel formation

entrapped enzymes might be reduced. Furthermore, enzymes may detach and slip out from the matrix over a period

These disadvantages can be overcome by immobilizing enzyme on the solid carrier backbone. Yagonia *et al.* (2014) covalently immobilized CALB to modified sol–gel matrix via hydroxyl groups after polycondensation of the matrix. The enzyme can avoid the extreme conditions of encapsulation because it is immobilized after the formation of sol–gel.

CALB is covalently immobilized on the surface of the support with its active site open to the substrate. Solvent accessible surface area (SASA) analysis indicates that CALB has three exposed lysine residues used to react and form covalent bonds with aldehydes of the support. The lysine residues are far from the active site to reduce conformational variations that can decrease the enzyme activity.

Increasing the amount of enzyme loaded per gram weight of the support and decreasing the support's particle size, with higher surface/volume ratio, can improve the activity of immobilized enzyme. CALB had higher activity on the sol–gel matrix with hydrophobic precursor.

6.3 Reactor System and Engineering Consideration

Once enzymes are immobilized, appropriate reactor type is to be selected for industrial reactions. In choosing a reactor system for a particular process, several factors are considered as follows:

- Cost to meet the product's specification (e.g., substrates, downstream processing)
- Kinetics of the reaction
- Chemical and physical properties of the immobilization support
- Methods for pH and temperature control
- Need of supplying and removing of gaseous components, if any and
- Stability of the enzyme

6.3.1 Reactor Types

6.3.1.1 Batch Reactors

These types of reactors consist of a tank having a stirrer and baffles inside to improve stirring efficiency and heat exchanger, if required. The enzyme and substrate have same residence times within the reactor. However, there may be a need for interim additions of enzymes and substrate in fed-batch operation. The product is removed after a fixed reaction time as rapidly as practically possible. Batch reactors are simple both in process development and operation. Thus, they are preferred for small-scale production with many different products. These reactors could be used for a number of different reactions. However, there are considerable periods for regular cleaning, refilling, and emptying which require much labor and services. This makes the operating cost of batch reactors higher than that of continuous processes. Batch-to-batch variations are inevitable for batch reactors.

6.3.1.2 Continuous Flow Reactors

Continuous flow reactor has also many advantages when using immobilized enzymes. Increased in productivity from same amount of enzyme can be achieved compared to that in batch processes, since the immobilized catalyst have higher residence time than the substrate within the reactor. These types of reactors offer constant reaction conditions thus more reproducible product can be obtained in quantity and quality. Continuous stirred tank reactor (CSTR), packed-bed reactor (PBR) and fluidized-bed reactor (FBR) are the examples of a continuous flow reactor.

Continuous Stirred Tank Reactor (CSTR). This reactor means a well-stirred tank containing the immobilized enzymeC inside. There is a continuous feeding of substrate, and at the same time continuous removal of the product is performed. In ideal case, there is complete backmixing and the product concentration in the effluent stream is identical with the concentration in the reactor. Since the substrate concentration is minimized due to backmixing, CSTRs are the preferred reactors for processes involving substrate inhibition. Control of temperature, pH, and the supply or removal of gases is easy. The shear from the mechanical stirring affects the immobilizing matrix material and thus limits the availability of the supports for the immobilized enzymes to prevent from easy disintegration.

Packed-Bed Reactor (PBR). Substrate stream flows at the same velocity, parallel to the reactor axis with no backmixing in an ideal case. All products emerge with the same residence time. The longitudinal position within the PBR is proportional to the time spent of the substrate and the product within the reactor. Consequently, the substrate concentration is maximum at the entry point of the reactor thus there is a possibility for substrate inhibition at the entry point. While product concentration becomes maximum at the outlet point, there is a possibility of strong product inhibition at the outlet of the reactor. PBRs are generally used with rigid immobilized catalysts, so it will not be compressed or distorted during the reaction with continuous flow. Particle deformation would result in reduced catalytic efficiency, poor mass transfer, and restriction to flow, which can lead to an increased pressure drop.

Fluidized-Bed Reactor (FBR). These reactors consist of a bed of immobilized enzyme which is fluidized by the upward flow of the substrate stream. The characters of FBR generally are between CSTRs and PBRs. The FBR is applied when fairly small immobilized enzyme particles are used in order to achieve a high catalytic surface area. Difficulties in scaling-up are the major drawbacks in the process development using FBR. Furthermore, changes in the flow rate of the substrate stream lead to changes in the flow pattern inside the reactor which in turn can cause unexpected results on the reaction.

Membrane Reactors. Semipermeable membrane where the free passage of the product molecules while retaining the enzyme within the membrane become possible is used for this reactor system. Therefore, there is no direct immobilization of enzymes. Membrane reactors can be applied in many cases as a continuous mode and allow an easy separation of the enzyme from the product. Soluble enzymes are generally used to avoid the cost and problems associated with other immobilization method. Membrane reactor systems are used for small-scale production. The major disadvantages of these reactors are the cost of the membranes and their need for frequent replacement or backwashing due to membrane fouling (Figs. 6.5 and 6.6).

6.3.2 Engineering Considerations

There are many issues when considering commercialization of immobilized enzyme system. (1) Immobilization yield: how much enzyme can be immobilized

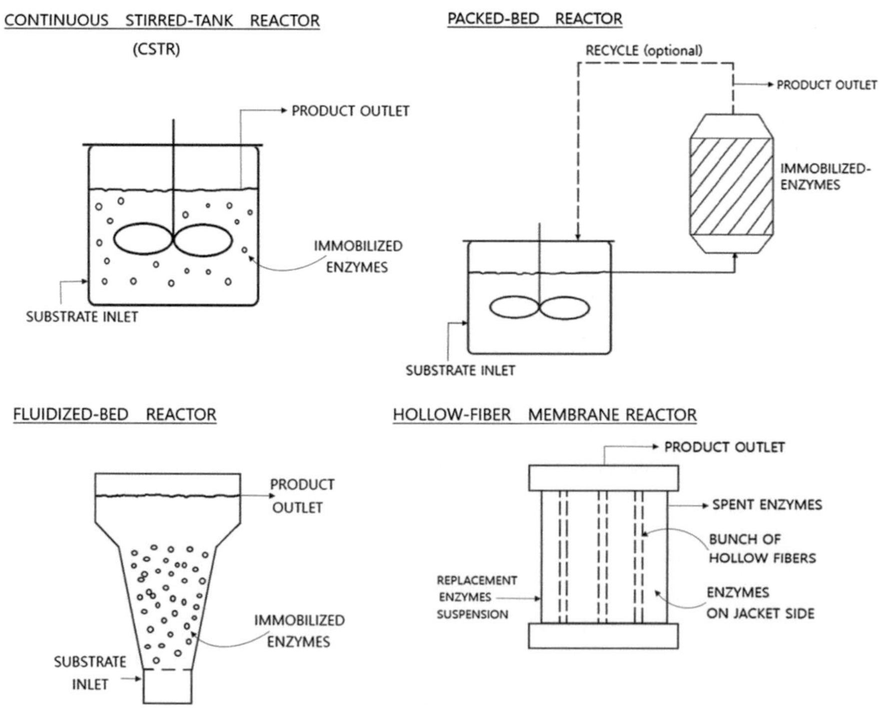

Fig. 6.5 Scheme of immobilized enzyme bioreactors

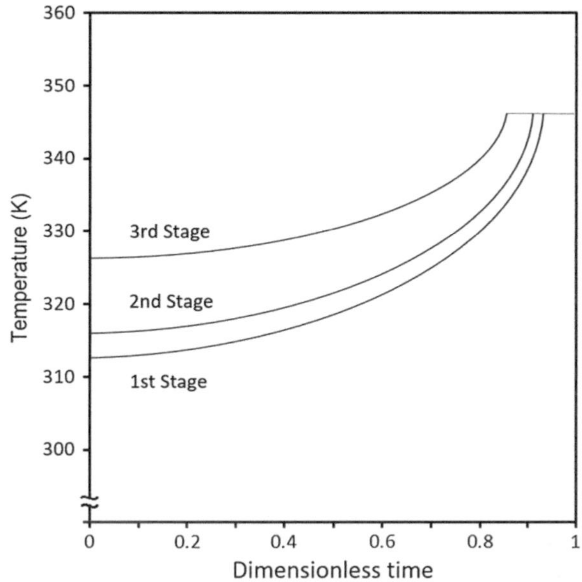

Fig. 6.6 Optimal temperature profiles for three-stage reactor system (Operating period = 200 h)

per gram of immobilization matrix? For commercial purpose, many matrices for immobilization of enzymes have been developed. Cost of the matrix and for immobilization is also important, (2) Enzyme reactor size, (3) Long term stability, (4) Contamination by microorganisms should be avoided or to be minimal, (5) Impurity treatment in raw materials is important. Impurities in raw materials may affect the reaction rate and impurities after the reaction should be removed, if required.

In 1960s, Tanabe Seiyaku in Japan commercialized L-aspartic acid production, which is important as raw material for sweetener Aspartame. It is known that polyacrylamide was used for immobilizing aspartase enzyme, where half-life of the immobilized enzyme at 37 °C was 120 days. By changing to carrageenan gel for the immobilization, half-life was increased to 680 days. This example shows the importance of the immobilization method, even though theoretical and molecular level mechanisms are not fully understood.

Engineering parameters are to be considered for scale-up of the reactor system: critical fluid velocity: fluid velocity affects shear rate which may weaken matrix, tensile strength is important when gas such as carbon dioxide is produced as a result of the reaction, external and/or internal mass transfer, operational stability (half-life), compression behavior, and pressure drop.

In industrial scale operation, if the cost of the enzyme is not expensive, free enzyme can be used instead of immobilized enzyme. Reuse of the immobilized enzyme is possible and can be considered. It depends on the cost of the immobilized enzyme. If the enzyme is expensive, the immobilized enzyme can be regenerated after the reaction. High temperature more than 60–70 °C can be employed instead of optimal temperature of the enzyme to avoid bacterial contamination. Instead, fast deactivation can be occurred. Half-life in real operation can be different from the half-life in lab operation, since many conditions are different such as consistency of the substrate, shear rate, etc. Stability during storage and shutdown should be considered. Legislative approval is to be considered for safety, when used especially for food additives. Nontoxic matrix for immobilization is to be used and the product safety is to be checked after the reaction. Even distribution of the flow to the reactor is important, especially for downward flow, since this affects pressure drop of the reactor and reactor performance.

Basic control parameters in immobilized enzyme reactor operations are temperature and pH. The reaction can be exothermic or endothermic and the pH of the solution can be changed as the reaction proceeds. It is not difficult to control these parameters in batch and continuous stirred tank reactors. However, it is not easy to control these parameters in packed-bed reactors which are the most widely employed commercially. If required, control of these parameters stagewise is used instead of single-stage operation.

More Issues

Another issue in immobilized enzyme reactor operation is the gradual decrease of the activity of the immobilized enzyme. This resulted in gradual decrease of the reaction rate and thus gradual decrease in the product concentration. As the enzyme activity decreases to a certain point such as half of the initial activity, then the reaction is stopped and used enzymes are replaced by new enzymes. Constant temperature or constant flow rate is desirable during the immobilized enzyme reactor operations. When considering gradual decrease of the activity, flow rate or temperature can be changed. For decreasing activity, flow rate can be thus decreased to keep the same reaction rate or the reaction temperature can be increased to compensate the decrease of the activity. In any case, the goal of the operation is to maximize the profit from the operation. For this purpose, optimal control of the reaction temperature in immobilized enzyme reactor was suggested (Yoon *et al.* 1989).

Reactivation of the immobilized enzyme was also suggested. After the reaction, the enzyme activity drops. If the cost of the enzyme is not high, used enzyme can be discarded. However, if the enzyme cost is very high, regeneration and reuse of the enzyme can be a good solution. First step of the regeneration is unfolding the irreversibly thermoinactivated enzyme using chemicals such as urea or guanidine chloride and splicing of the S–S bonds using mercaptoethanol or other chemicals. After unfolding, refolding step is required by reoxidation of S–S bonds or other reactions to make native enzyme conformation.

In situ separation of product to reduce the product inhibition and to recover the product at the same time is also important. For this purpose, many methods have been suggested such as aqueous 2-phase reaction and 2-phase reaction using organic solvent system.

Also instead of single-stage reactor operation, multistage reactor operation and recycled reactor operation can be considered for efficient and economic operation. The details on the methods on reactor design and operation, specific reaction engineering books and references can be a good guide.

Case Study: Production of High Fructose Corn Syrup (HFCS)

Glucose can be converted to glucose–fructose mixture using glucose isomerases (EC 5.3.1.5). Glucose isomerase was initially known to convert xylose into xylulose, so-called xylose isomerase but later called as glucose isomerase since it also has an affinity toward glucose. Since it has an activity toward glucose, the enzyme is being used for the production of glucose–fructose mixture, a sweet alternative to sucrose. Glucose isomerase could convert glucose partially into fructose because of reversible reaction. Since the source of glucose is in many cases glucose from corn and the fructose concentration in the product is high, the product is often called high fructose corn syrup. The sweetening power of glucose–fructose mixture is almost the same as that of sucrose.

Clinton Corn Processing Company at the end of 1960s produced fructose from glucose at a large scale. Before the development of enzymatic isomerization, chemical isomerization process utilizing sulfuric acid was used, where decomposition and production of harmful by-products and brownish coloring appearance were occured. Since the enzyme was expensive, it was challenging to make the production of fructose syrup using enzymes. This was addressed by immobilizing thermostable glucose isomerases in packed-bed reactor under 65 °C—high temperature is preferred to prevent contamination from the microorganisms as shown in Fig. 6.7. 42% fructose mixture was produced on 100 ton per day in three 2.2 m^3 columns. The mixture was therefore concentrated up to 55% (55% fructose: 45% glucose, HFCS 55) by chromatography or other separation tools to meet the same

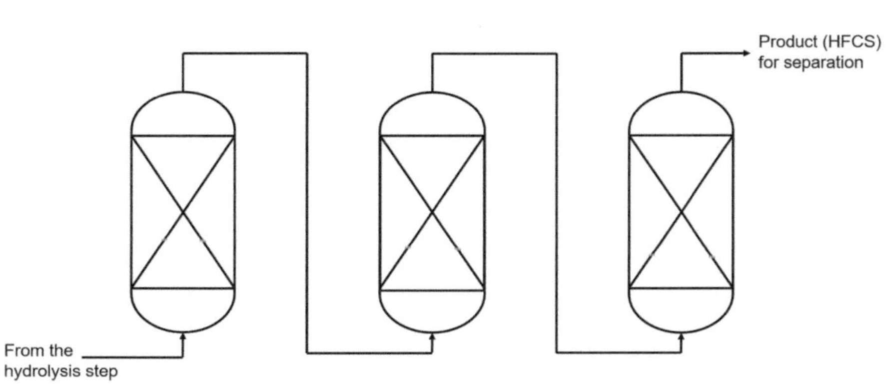

Fig. 6.7 HFCS production reactors employing thermostable glucose isomerase. The reactor is packed with immobilized enzymes

sweetness level with sucrose. The enzyme was initially immobilized using free-form enzyme. Later, immobilization using whole cells was also introduced not to infringe the patent issue, where permeabilization important to allow the substrate can penetrate into the cells and to react with the enzymes in the cell and then the product diffuse out.

Further Discussion

1. How site-specific immobilization of enzyme can be done to allow for the active site open to substrate?
2. For substrate inhibition kinetics, what reactor type for immobilized enzyme is better and why? For product inhibition kinetics, what reactor type for immobilized enzyme is better and why?
3. Enzymes in immobilized form also become deactivated. How can immobilized enzyme system be regenerated? Can the spent immobilized enzyme be continuously replaced to new immobilized enzyme?
4. Sometimes whole cells are immobilized instead of free enzymes for industrial applications. What is the background of immobilizing whole cells and how to immobilize?

References

Bruns N and Tiller JC. Amphiphilic network as nanoreactor for enzymes in organic solvents. *Nano Letters*, 2005, 5:45–48.

Hernandez K and Fernandez-Lafuente R. Control of protein immobilization: Coupling immobilization and site-directed mutagenesis to improve biocatalyst or biosensor performance. *Enzyme and Microbial Technology*, 2011, 48:107–122.

Kim J, Grate JW. Single-enzyme nanoparticles armored by a nanometer-sacle organic/inorganic network. *Nano Lett.*, 2003, 3:1219–1222.

Min K and Yoo YJ. Recent progress in nanobiocatalysis for enzyme immobilization and its application. *Biotechnology and Bioprocess Engineering*, 2014, 19:553–567.

Palomo JM, Fernandez-Lorente G, Mateo C, Ortiz C, Fernandez-Lafuente R and Guisan JM. Modulation of the enantioselectivity of lipases via controlled immobilization and medium engineering: Hydrolytic resolution of mandelic acid esters. *Enzyme and Microbial Technology*, 2002, 31:775–783.

Regan DL, Lilly MD and Dunnill P. Influence of intraparticle diffusional limitation on the observedkinetics of immobilized enzymes and on catalyst design. *Biotechnology and Bioengineering*, 1974,16:1081–1093.

Reetz MT, Zonta A and Simpelkamp J. Efficient heterogeneous biocatalysts by entrapment of lipases in hydrophobic sol-gel materials. *Angewandte Chemie*, 107:373–376; *Angewandte Chemie International Edition in English*, 1995, 34:301–303.

Reetz MT. Entrapment of biocatalysts in hydrophobic sol-gel materials for use in organic chemistry. *Advanced Materials*, 1997, 9:943–954.

Sheldon RA. Enzyme immobilization: the quest for optimum performance. *Adv. Synth. Catal.*, 2007, 349:1289–1307.

Tischer W and Wedekind F. Immobilized enzymes: methods and applications. *Topics in Current Chemistry*, 1999, 200:95–126.

Yoon SK, Yoo YJ, and Rhee H. Optimal temperature control in a multi-stage immobilized enzyme reactor system. *J. Fermentation and Bioengineering,* 1989, 68:136–140.

Yagonia CFJ, Park K and Yoo YJ. Immobilization of *Candida antarctica* lipase B on the surface of modified sol-gel matrix. *Journal of Sol-Gel Science and Technology*, 2014, 69:564–570.

Zhang B, Weng Y, Xu H and Mao Z. Enzyme immobilization for biodiesel production. *Applied Microbiology and Biotechnology*, 2012, 93:61–70.

Chapter 7
Enzymes in Non-conventional Media

7.1 Enzymes in Non-conventional Media

Traditionally enzyme reactions have been in many cases performed in aqueous buffer systems. Since the environments inside the cells are rather hydrophilic in some part and hydrophobic in other part of the cells, enzyme reactions can be also performed in hydrophobic condition. Historically, organic solvent was used for steroid bioconversion. Since solubilities of substrate and product steroids are very low in water, they can be solubilized using organic solvent system. First, cells were cultivated to synthesize the enzyme for steroid biotransformation; the cell wall was then permeabilized; and finally two-phase reaction was performed to produce the steroid product (Carrea *et al.* 1988) (Fig. 7.1).

Non-conventional media refers to systems that employ solvents other than water. Solubility of organic solvents in water, depending upon the miscibility, results in many variations in reaction scheme. Enzyme reactions in solvent systems can be classified into three categories

(1) organic–aqueous biphasic systems (water-immiscible organic solvent),
(2) co-solvent systems (water-miscible organic solvent),
(3) enzymes in nearly anhydrous solvents (low water content, water-immiscible organic solvent).

7.1.1 *Enzymes in Organic Solvents*

Enzyme reactions near anhydrous condition have been extensively studied since 1980s. Minimal water layer and pH memory effect were investigated to explain

Y.J. Yoo *et al.*, *Fundamentals of Enzyme Engineering*,
DOI 10.1007/978-94-024-1026-6_7

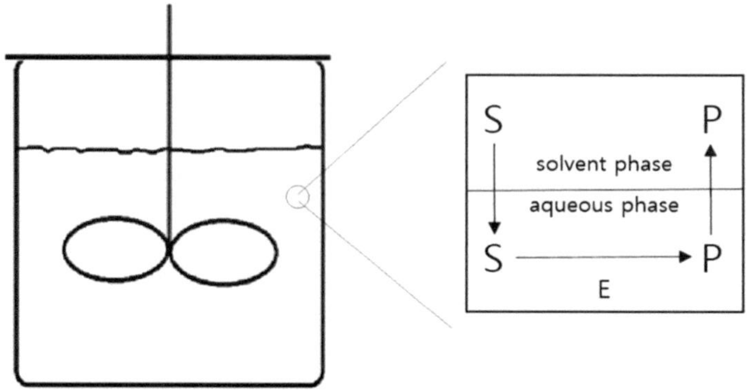

Fig. 7.1 Scheme for 2-phase enzyme reaction

the behavior of enzymes and relationship between enzyme activity and solvent property, but not enough so far in explaining.

Enzymes in organic solvents have many advantages over the traditional aqueous bioconversions by showing different properties: (1) changes in the enantioselectivity, (2) reversal of the thermodynamic equilibrium (e.g., synthesis of peptides from amino acids), and (3) inhibition of water-mediated side reactions (e.g., hydrolysis of acid anhydrides and acyl halides) (Castro and Knubovets 2003). The organic media is favored over aqueous one when substrates dissolve better in the organic solvent. Organic solvents reduce the risk of microbial contamination. Easy recovery and reuse of enzymes are possible even without immobilization. However, there are also drawbacks with enzymes in organic solvents such as reduced stability and low reaction rate.

Various strategies to overcome the disadvantages of organic solvent systems, especially in terms of improving the enzyme activity and stability, include the following: solvent engineering; enzyme engineering; covalent attachment of amphipathic compounds (e.g., PEG, aldehydes, and imidoesters); non-covalent interactions with lipids or surfactants; entrapment in water–oil microemulsions or reverse micelles; immobilization; and utilization of lyophilized enzyme powders or cross-linked crystals.

pH memory of enzyme in organic solvent (Tao and Klibanov 1998) describes that enzyme molecules seem to "remember" the aqueous conditions from which they were prepared. When lyophilized enzymes were suspended in organic solvent, very low specific activities were observed in many cases. However, if the enzymes were dissolved in water and then precipitated in anhydrous organic solvents, activities were shown to be much greater. Thus, the pH of an enzyme-dissolved solution must be adjusted to an optimum prior to dehydration.

Deactivation of enzyme activity was also observed during lyophilization. A possible prevention of the enzyme inactivation can be achieved by adding lyoprotectants such as polyethylene glycol, a polyol, which protect the enzyme conformation during dehydration. Another approach is to add phenolic and aniline substrates that bind to the hydrophobic active site pocket.

7.1.2 The Role of Solvent Properties

Organic solvent's ability to affect the properties of enzymes such as variable selectivities is a major advantage. The term "medium engineering" was coined in this regard. Researches to predict the properties of an enzyme depending on the parameters of an organic solvent are carried out. The parameters include dielectric constant, dipole moment, polarization constant, and most of all, log *P* value. Log *P* is the water/octanol partition coefficient, and defined as follows, where *A* is a dilute solution below the solubility limit:

$$\log P = \log \frac{[A]_{\text{water}}}{[A]_{n\text{-octanol}}} \tag{7.1}$$

n-Octanol is used as a reference because its hydrophobic tail and hydrophilic head resemble those of phospholipid in biological membranes. Log *P* values, boiling points, and enthalpies of evaporation for frequently used solvents are listed in Table 7.1.

Table 7.1 Log *P* values, boiling points, and enthalpies of evaporation for solvents

Solvent	Log *P*	B.P. (°C)	ΔH_{vap} (kJ/mol)
1,4-dioxane	−1.1	101.6	35.8
N,N-dimethylformamide	−1.0	153.0	60.5
Methanol	−0.76	65.0	35.4
Acetonitrile	−0.33	82.0	32.7
Ethanol	−0.24	78.5	40.5
Acetone	−0.23	56.2	32.0
Tetrahydrofuran	0.49	67.0	28.8
Ethyl acetate	0.68	77.1	34.8
n-butanol	0.80	108	44.0
Diethyl ether	0.85	34.6	26.6
Methyl tert-butyl ester	1.15	55.2	29.7
Chloroform	2.0	61.1	29.7
Benzene	2.0	80.1	30.8
Toluene	2.5	110.6	39.2
Cyclohexane	3.2	80.7	32.8
Octane	4.5	125.3	34.6
Dodecanol	5.0	259	–
Dioctyl phthalate	9.6	–	–
Water	–	100	40.7

Table 7.2 Dependence of enantioselectivity E of subtilisin on solvent properties in the transesterification reaction

Solvent	Dielectric constant ε	Log *P*	Enantioselectivity *E*
Dioxane	2.2	−1.1	61
Benzene	2.3	2.0	54
THF[a]	7.6	0.49	40
Pyridine	12.9	0.65	31
DMF[a]	36.7	−1.0	9
Nitromethane	35.9	0.35	5
Acetonitrile	35.9	−0.33	3
MAA[a]	191.3	–	3

[a]*THF* Tetrahydrofuran, *DMF* dimethylformamide, *MAA* metacrylamide

Table 7.3 Asymmetric hydrolysis of nifedipine diester

R	Solvent	Configuration	ee (%)
$tBuC(O)OCH_2$	Cyclohexane	*R*	88.8
$tBuC(O)OCH_2$	Diisopropyl ether	*S*	>99
$EtC(O)OCH_2$	Cyclohexane	*R*	91.4
$EtC(O)OCH_2$	Isopropyl ether	*S*	68.1

The enantioselectivity of the transesterification of vinyl butyrate with sec-phenylethanol was greatly influenced by the solvent. However, it was the dielectric constant of the solvent, not log *P*, which correlated the best with the *E* values as shown in Table 7.2.

Interestingly, enantioselectivity of enzyme can be remarkably increased in organic solvent. Strikingly even enantioselectivity was inversed in some organic solvents. Nifedipines have been used in cardiovascular therapy as calcium antagonists. *Pseudomonas* sp. lipase catalyzed hydrolysis of methylene-oxypropiony or -pivaloyl diesters. Totally reversed enantioselectivity was found in organic solvents (Table 7.3).

7.1.3 The Role of Water Molecules on Enzyme Activity and Stability

Quite often, drastic decrease of enzyme activity has been observed in organic solvents compared with in water, which may be a main disadvantage of applying organic solvents for biocatalysis. Elevation of reaction temperature can compensate the disadvantage by recovering enzyme activity to some extent since high temperature can increase flexibility of enzyme molecules. However, minimum content of water contained in enzyme molecule can influence significantly on enzyme activity and stability. Water concentration is expressed quantitatively as

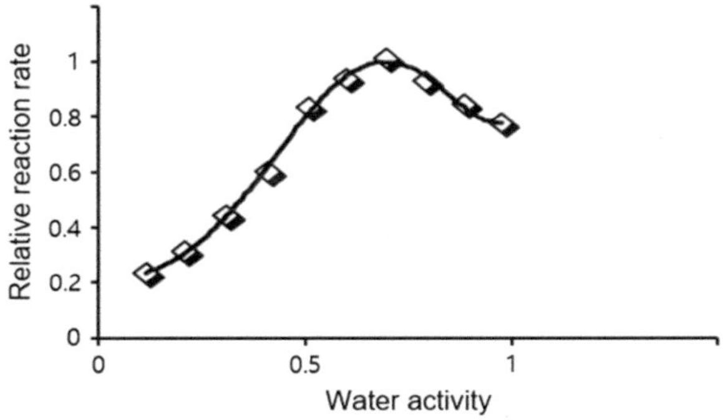

Fig. 7.2 Typical profile showing the relationship between reaction rate and water activity

water activity a_W, which is the product of the concentration of water and water activity coefficient. Figure 7.2 shows typical example of relationship between reaction rate and water activity.

In most cases enzyme activities increase up to a limit with gradual hydration of enzyme. In more hydrophilic solvents, the enzyme and the solvent compete for water, hence leading to more required water to reach the maximum activity. Almost identical patterns for enzyme activities based on the water content of enzyme were observed and about 1000 H_2O molecules are known necessary to hydrate an enzyme by monolayer, which is corresponding with the maximal activity. Excess of water causes a decline of activity since the formation of a second hydration layer may act as a barrier to mass transport. In case of lipase-catalyzed reaction in organic solvents, optimal water activity was reported as a_W of 0.55 irrespective of solvent types, underscoring the importance of water activity of enzyme. Therefore, the water activity of the enzyme needs to be controlled for optimal reaction using saturated salt solutions. For example, 0.113 a_W and 0.936 a_W of enzyme can be adjusted using LiCl and KNO_3, respectively.

7.1.4 Other Non-conventional Media

Modification with Polyethylene Glycols (PEGs)

Reactive groups on the surface of an enzyme, such as lysine ε-amino group, and amphipathic polymers such as PEG can form covalent bonds. The modification involves the following stages: (1) the polymer's functional group activation to increase activity; (2a) transient intermediate formed between the activated functional group and the enzyme; (2b) formation of a stable enzyme–polymer complex from the intermediate; (3) purification of the complex from unmodified enzymes (see Fig. 7.3).

Polyethylene glycol is the most extensively studied polymer. "Pegylation" is the term applied for the modification process using PEG. *p*-nitrophenyl chloroformate, *N,N*-disuccinimidyl carbonate and cyanuric chloride are frequently used activators in pegylation of proteins, leading to significant improvements in their functions.

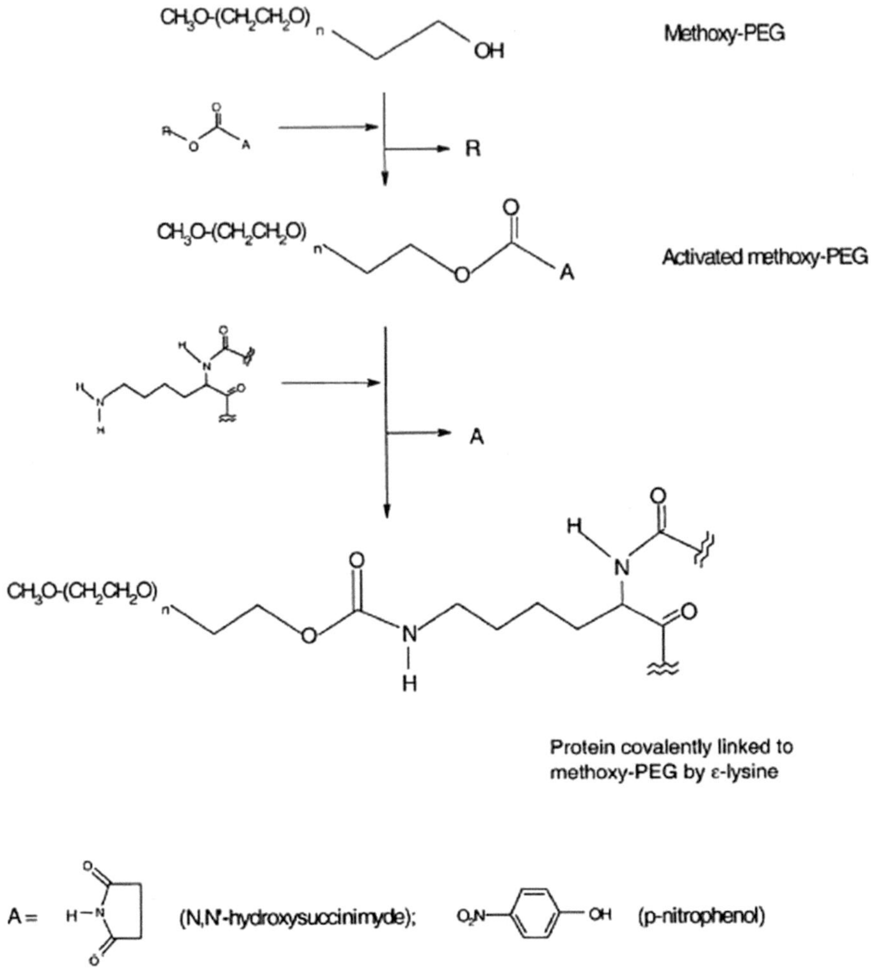

Fig. 7.3 Modification of a protein with methoxypolyethylene glycol (Castro and Knubovets 2003)

Reverse Micelles

Reverse micelles are surfactant layers covering a pool of water, of which the hydrophobic group interacts the bulk solvent as shown in Fig. 7.4. This type of heterogeneous solvent system is similar to the natural environment in the cell, and thus it is attractive for biocatalysis. The sodium bis-2-ethylhexyl sulfosuccinate (AOT)/isooctane system is very stable without co-surfactants and frequently used for this purpose.

The selection of the reversed micelle system for an enzyme is typically based on the successes reported in the literature for similar enzymes and reactions due to complexity. Before its application at industrial scale, there are some issues to be solved. Industrial-scale product recovery is difficult due to the surfactants of the

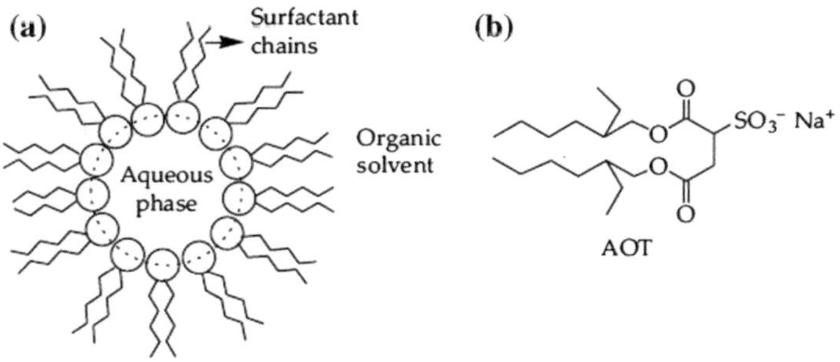

Fig. 7.4 Scheme of a reverse micelle and chemical structure of AOT (Krieger *et al.* 2004)

system. The use of ultrafiltration membrane could be a possible solution because it can retain the micelles and the enzyme while the small molecules of substrates and products pass freely. However, for continuous process using membrane reactors, the enzyme needs to be stable for long periods.

Supercritical Fluids

Supercritical fluids, which means a state between the gaseous and liquid phases of the compound, have hydrophobic solvent-like properties. Near-supercritical fluids such as carbon dioxide, freons, hydrocarbons (i.e., ethane, ethane, and propane) or inorganic compounds (SF_6 and N_2O) are exploited to create supercritical solvent systems. Supercritical carbon dioxide ($scCO_2$) is the most commonly used fluid since it is non-toxic and can be removed easily after the reaction. The application of supercritical fluids increases diffusion rates which can facilitate transport phenomena and increase the bioconversion rate. However, the main drawback of supercritical reaction media is the necessity of high pressure-resistant reactors and auxiliary which increases production cost.

Ionic Liquids (ILs)

Ionic liquids are liquid-phase organic salts where their properties can be tuned to modify enzyme activity and facilitate product recovery. Ionic liquids are claimed to contribute to green chemistry since they are nonvolatile at room temperature, in contrast to organic solvents.

7.2 Stability in Organic Solvent

Many technologies have been developed to prevent the enzyme deactivation in organic solvent system, such as solvent engineering, immobilization of enzyme, chemical modification of enzyme including PEglation, and protein engineering.

Simple methods for keeping enzymes active in organic solvents involved lyophilization or chemical modification. The enzymes can be potentially damaged

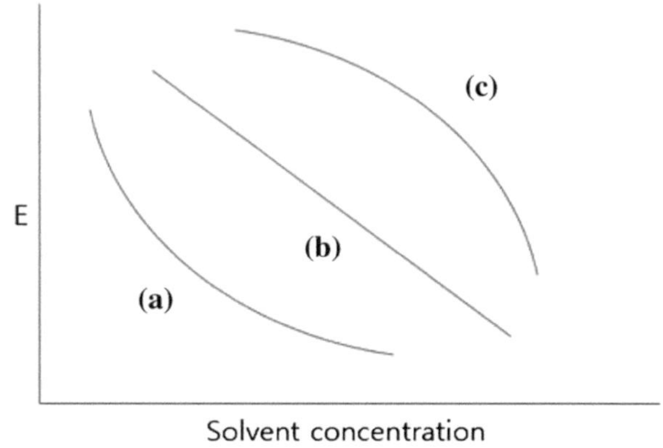

Fig. 7.5 Enzyme activity dependence on solvent **a** water, glycerol, EG, formamide **b** methyl formamide, dimethyl formate **c** methanol, ethanol, toluene

during lyophilization conditions, which affect their activity and other properties in organic medium.

On the other hand, chemical modification of enzyme, at the surfaces by introducing hydrophobic moieties to lysine residues, gives promising opportunity for enhancing solubility, stability, and substrate preference in organic solvents. The attachment of phthalic anhydride to horseradish peroxidase (HRP) increased the enzyme's organic solvent stability twice and also led to an increase in thermal stability four times (O'Brien and Smith 2003).

Solvent selection is a way to avoid solvent effect of the enzyme activity. Since it is important not to destruct hydration shell of enzyme by solvent, solvent selection is an important issue.

Since polyols have high viscosity which is not good for mixing and low polarity which gives low substrate solubility, polyols are not widely used for enzyme reaction as reaction media (Fig. 7.5).

7.2.1 Molecular Understanding

One approach is to find stable enzyme and analyze the characteristics in molecular level. For example, thermostable alcohol dehydrogenase which can be used for asymmetric reduction of ketones in organic solvent was found and characterized (Hoellrigl *et al.* 2008). Mutation techniques such as directed evolution and site-directed mutagenesis can be applied to alter enzyme specificity and enhance its activity in organic solvent. Directed evolution studies show that after two generations of mutations and activity screening, the activity of cytochrome P450 BM-3 in organic co-solvent mixture increased tenfold over that of the wild type (Wong *et al.* 2004). By mutation of the residue, residues might prevent solvent access to the heme and reductase domains and act to maintain the enzyme domains in their proper orientation and geometry.

The analysis on the mutants having high activity in organic solvent made by directed evolution showed that hydrogen bonds were strengthened. The result was successfully applied to increase the organic solvent stability of lipase, of which approach is widely used for enzymes in organic solvent system (Park *et al.* 2012).

Another approach is to search the penetration site of organic solvent molecules to enzymes. Ethanol molecules which penetrated into the lipase A enzyme were found by molecular simulation, which were considered to destabilize the conformation of the enzyme by affecting hydrophobic interaction and hydrogen bonding. By changing the amino acids which were affected by ethanol to bigger ones to prevent the ethanol penetration into the enzyme, organic solvent stability was improved (Park *et al.* 2013).

Interaction between enzymes and solvent was investigated. Structural motion of the enzyme lipase B in methanol was studied using molecular dynamics (MD) simulation. The surface residues of CALB which have higher root mean square deviation (RMSD) were considered to affect the enzyme motion. In general, overall RMSD in organic solvent is less than the overall RMSD in aqueous solvent. However, local RMSD changes are different depending on the particular amino acid residues of the enzyme. Thus the amino acid residues greatly affected by organic solvent were selected as candidate for mutations. By adopting computational design to lower the RMSD, the mutant enzymes which showed improved stability in organic solvent was made (Park *et al.* 2014).

Further Discussion

1. Hydrolase enzymes such as protease, lipase, and amylase show different reaction schemes with or without solvent system. Find examples.
2. Why enzyme's characteristics are changed in the presence of solvent, for example, polymerization occurs under solvent condition?
3. Solvents in many cases are toxic and flammable. How to prevent such potential problems in industry?
4. Water activity was suggested to explain the enzyme activity in organic solvent system. What rationale was used for the explanation?
5. What enzyme engineering methods have been suggested so far to maintain or increase the enzyme activity and to prevent the enzyme from deactivation in organic solvent system?

References

Castro GR and Knubovets T. Homogeneous biocatalysis in organic solvents and water-organic mixtures. *Critical Reviews in Biotechnology*, 2003, 23(3):195–231.

Carrea G, riva S, Bovara R and Pasta P. Enzymatic oxidoreduction of steroids in two-phase systems: effects of organic solvents on enzyme kinetics and evaluation of the performance of different reactors, *Enzyme and Microbial Technology*, 1988, 10:333–340.

Hoellrigl V, Hollmann F, Kleeb AC, Buehler K and Schmid A. TADH, the thermostable alcohol dehydrogenase from *Thermus* sp. ATN1: a versatile new biocatalyst for organic synthesis. *Appl. Microbiol. Biotechnol.*, 2008, 81:263–273'.

Krieger N, Bhatnagar T, Baratti JC, Baron AM, de Lima VM and Mitchell D. Non-aqueous biocatalysis in heterogeneous solvent systems. *Food Technology and Biotechnol*ogy, 2004, 42:279–286.

O'Brien AM, Smith AT and Fagain CO. Effects of phthalic anhydride modification on horseradish peroxidase stability and activity. *Biotechnology and Bioengineering,* 2003, 81:233–240.

Park HJ, Joo JC, Park KM and Yoo YJ. Stabilization of *Candida antarctica* lipase B in hydrophilic organic solvent by rational design of hydrogen bond. *Biotechnology and Bioprocess Eng.,* 2012, 17:722–728.

Park HJ, Joo JC, Park KM, Kim YH and Yoo YJ. Prediction of the solvent affecting site and the computational design of stable *Candida antarctica* lipase B in a hydrophilic organic solvent. *J of Biotechnology*, 2013, 163:346–352.

Park HJ, Park KM, Kim YH and Yoo YJ. Computational approach for designing thermostable *Candida antarctica* lipase B by molecular dynamics simulation. *J of Biotechnology,* 2014, 192:66–70.

Tao K and Klibanov AM. On enzymatic activity in organic solvents as a function of enzyme history. *Biotechnol. Bioeng*., 1998, 57:746–750.

Wong TS, Arnold FH and Schwaneberg U. Laboratory evolution of cytochrome P450BM-3 monooxygenase for organic cosolvents. *Biotechnology and Bioengineering*, 2004, 85:351–358.

Part III
Molecular Understanding of Enzymes

Chapter 8
Engineering Tools for Enzymes

Since new millennia, application of industrial enzymes in manufacturing process including pharmaceuticals, fine chemicals, bio-based chemicals remarkably gained much attention. There is a fundamental limitation to apply industrial enzymes since enzymes have not been evolved to meet the requirements as biocatalysts. Enzyme engineering can be used to satisfy the required performances of industrially applicable biocatalysts. To satisfy the critical demands as biocatalysts, enzymes must be evolved to show high specific selectivity as well as tolerances against high temperature and harsh pH conditions.

Site-directed mutagenesis was developed by Michael Smith in 1978 and enzyme properties were improved through error-prone PCR by Frances H. Arnold in 1993, and DNA shuffling was reported by W. P. C. Stemmer in 1994. Since then, various tools for enzyme engineering have emerged for two decades. Rational design based on structural information of enzyme and random approach including directed evolution not requiring 3-D structural information of enzyme may represent key enzyme engineering methods. A general scheme for these representing methods is shown in Fig. 8.1.

Evolved substitution of amino acid residues in an enzyme and subsequential screening of positive clones showing desired traits are the basic procedure of directed evolution. On the while, rational design is dependent on the structural information to determine the target amino acid residues to substitute with proper ones. Directed evolution and rational design can be applied in tandem for better performances of enzymes applicable in an industrial scale.

Y.J. Yoo *et al.*, *Fundamentals of Enzyme Engineering*,
DOI 10.1007/978-94-024-1026-6_8

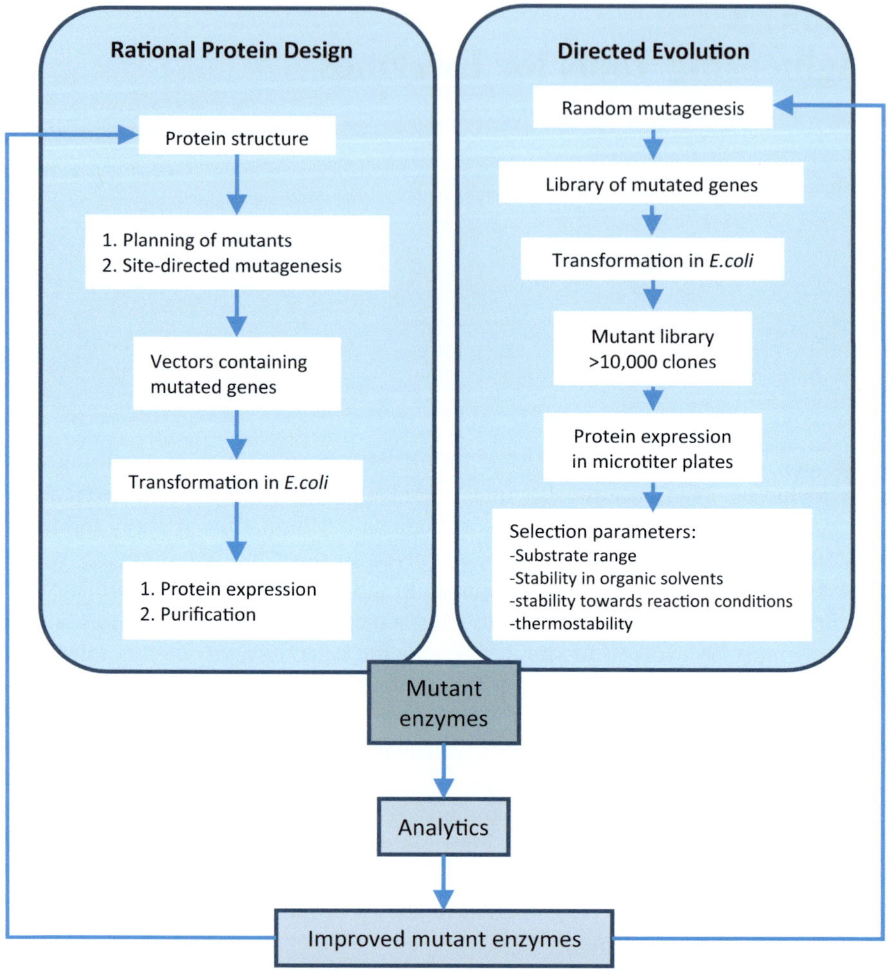

Fig. 8.1 Comparison of directed evolution and rational design when using *E. coli* as a host strain (Bornscheuer 2001)

8.1 Random Mutagenesis

Before the development of protein engineering, random mutagenesis was an only method to get the mutants from the microoranisms. Natural selection, originally suggested by Darwin is a conceptual basis of directed evolution. Directed evolution is useful not only for enzymes of known structure, but also for those whose structure–function relationship is not well understood. Stemmer and Arnold suggested directed evolution technology in the mid- and late 1990s. They invented molecular biology methods through which biocatalysts can be improved via an

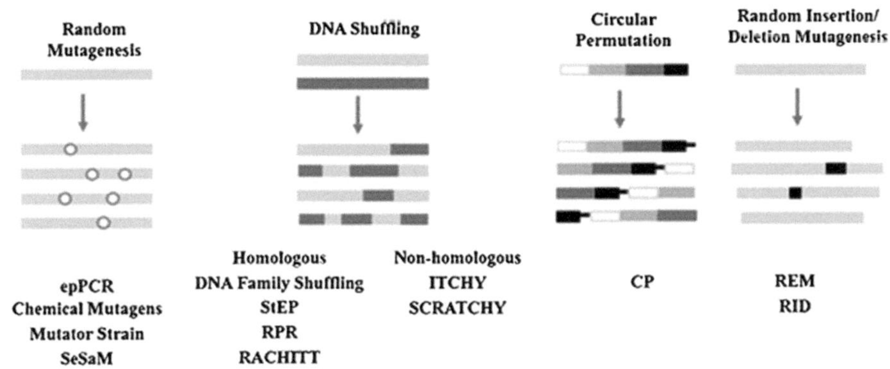

Fig. 8.2 Comparison of the strategies for directed evolution

in vitro version. The creation of molecular diversity is the cornerstone of directed evolution. The generation of DNA sequence diversity is one of key steps for the successful directed evolution and many DNA manipulation techniques have been suggested to generate diverse DNA sequences with more feasibility. A wide range of strategies are now available for creating genetic diversity for directed evolution with different characteristics (Fig. 8.2).

8.1.1 Random Mutagenesis for Genes

Random mutagenesis is the simple and easy method for generating DNA sequence diversities. Discovery of enzyme mutants showing desired traits can be much effectively performed when combined with high-throughput screening (HTS). Random mutagenesis of specific genes can be achieved by error-prone PCR or by sequence saturation mutagenesis. The experimental process of directed evolution using random mutagenesis is shown in Fig. 8.3. The comparison of experimental process of the main strategies for random mutagenesis is shown in Fig. 8.4.

DNA sequence is diversified by error-prone PCR (epPCR). Dramatic elevation of error rate of *Taq* DNA polymerase lacking of $3' \rightarrow 5'$ exonuclease proofreading function is fundamental basis for epPCR. For example, non-optimal conditions for DNA polymerization including high concentration of $Mg^{2+}$and Mn^{2+}ions or the unbalanced dNTP concentrations, can result in the elevation of the error frequency of common DNA polymerase from 0.001 to 1%. However, most DNA fragments generated through epPCR failed to be cloned during the subcloning procedures, being a bottleneck in the construction of the mutant library. Wild enzyme coding gene is ligated with a plasmid as first step and then the plasmid ligated with target

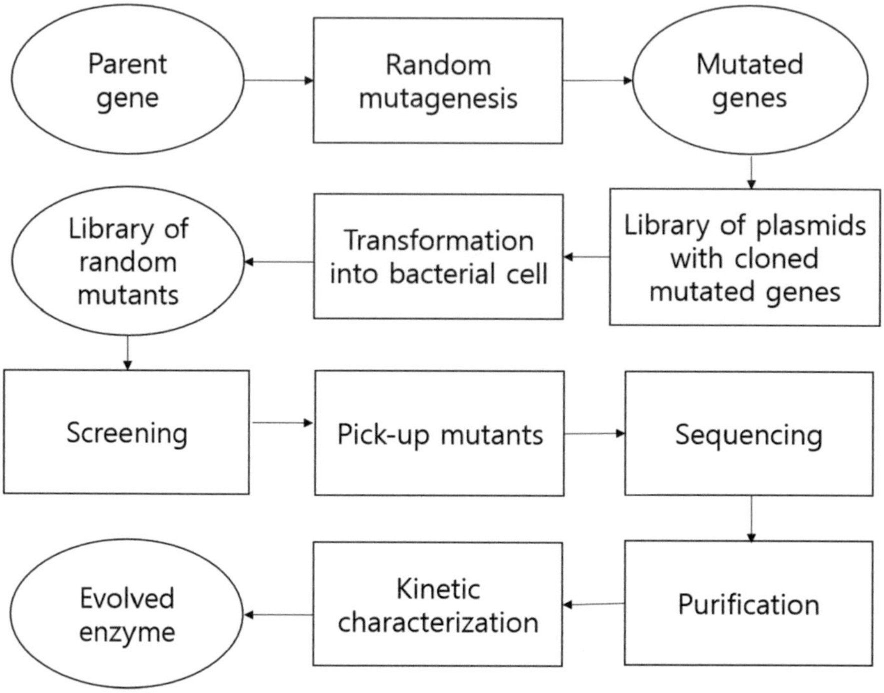

Fig. 8.3 Main experimental steps for random mutagenesis (Labrou 2010)

gene is polymerized by Taq DNA polymerase under error-prone conditions to generate diverse DNA sequences, followed by screening procedures (Fig. 8.4).

Mutator strains deficient of DNA repair machinery make it easy and simple to introduce point mutations randomly. In 1996, Greener developed mutator strain showing 5000 time higher mutation rate compared with wild one. This strain *E. coli* XL1-Red was constructed by knock-out of three DNA repair systems, MutS, MutD, and MutT (Greener 1996).

Sequence saturation mutagenesis (SeSaM) was suggested to replace epPCR (Wong 2004). Random substitution for a target DNA at the level of every single base can be made in SeSaM. SeSaM is a four-step patented method which represents a breakthrough in directed evolution and can be accomplished within 2–3 days.

Chemical mutagenesis has a long history and is regarded as DNA mutation method acceptable mainly for food industry, because it does not involve the introduction of heterologous gene sequences. Since 1994, several chemicals have been used to introduce random mutations into a gene of interest in vitro, including ethyl methane sulfonate (EMS), nitrous acid (HNO_2), hydroxylamine, bisulphate, methoxylamine, and so on. The key merits of chemical mutagenesis are simplicity and low cost, and the main disadvantage is the lack of control of mutation rate and limited amino acid substitutions.

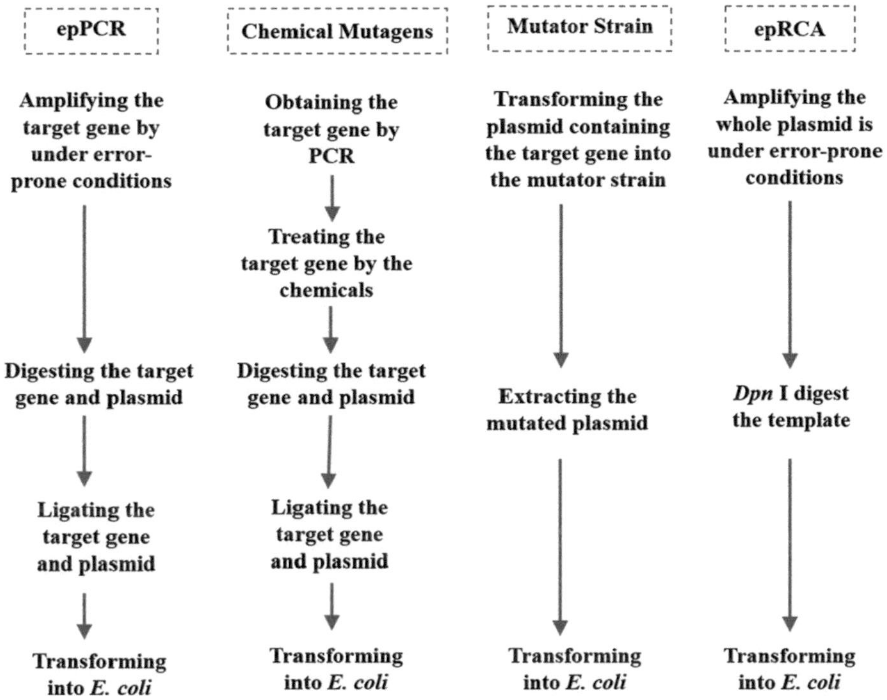

Fig. 8.4 Comparison of experimental process of the random mutagenesis strategies when using *E. coli* as a host strain

8.1.2 Molecular Breeding of DNA

Homologous DNA Shuffling

In 1994, Stemmer presented the DNA shuffling technique in the Nature (Stemmer 1994a) and PNAS (Stemmer 1994b), as the first homologous recombination method, and built a theoretical basis for developing new DNA shuffling methods. DNA shuffling commonly consists of the random digestion of parent genes by DNases into DNA fragments, and subsequent reassembly of the DNA fragments into a full-length gene by primerless PCR (self-priming PCR): sequence homology found on DNA fragments results in recombined annealed genes through template switch (Fig. 8.5). DNA shuffling result in not only molecular diversity by recombination but also combined effective mutations from source genes. Reassembly process of DNA fragments into full-length gene through self-priming PCR makes the main difference distinguished from other random point mutagenesis.

Improved enzyme will be evolved successfully through DNA shuffling followed by appropriate screening method. Classical DNA shuffling started from single gene and generated various diverse genes including randomized point mutations. The evolution to enzyme showing a desired trait will be inefficient since the

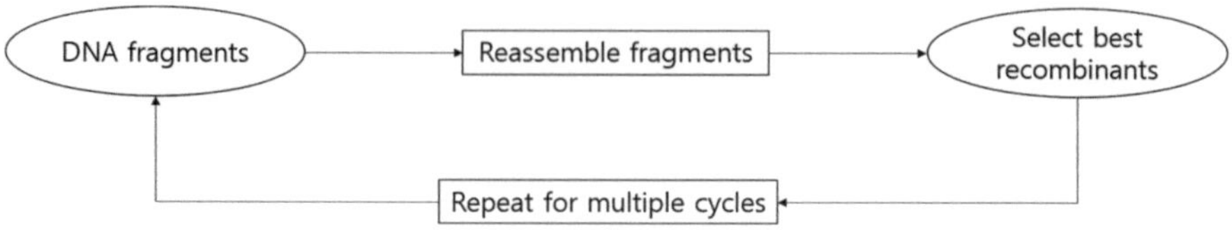

Fig. 8.5 Main experimental process of DNA shuffling

accumulation of beneficial mutation is less compared with deleterious or neutral point mutations. In 1998, Stemmer presented a new tool DNA family shuffling (Crameri 1998). This technique utilizes native homologous genes from diverse species rather than in vitro-generated mutants as the driving force for directed evolution. Reassembly of the randomized fragments using self-priming will result in the generation of newly diversified chimeric genes.

However, as one of the most powerful techniques for enzyme engineering, DNA family shuffling is not flawless and always includes the weak points caused by reassembling process (unshuffled molecules), because the formation of chimeric sequences is prevented. Single-stranded DNAs (ssDNAs) as the template instead of dsDNAs were used for the increased efficiency to hybrid formation. DNA shuffling can be improved using the endonuclease V to random DNA fragmentation instead of DNase I.

In 1998, Arnold's group reported a new effective protocol method so called as staggered extension process (StEP) (Zhao 1998). StEP consists of using the template sequences as self-primers, which subsequently performed through repeated cyclic denaturation and short annealing/elongation. Unlike classical DNA shuffling generating random fragments by digestion, StEP uses elongation for further recombination by using primers to produce fragments.

In 1998, Arnold's group presented a novel protocol, random-priming recombination (RPR) (Shao 1998), as a good surrogate. Randomized DNA oligomers are utilized to produce a lot of short fragments paired with different regions of the template DNA. These small fragments can act as primer each other due to close homology. This will result in the full-length genes through further reassembly process by PCR procedures. The method RACHITT employs no PCR procedures, but short sized parental genes will be hybridized. RACHITT can increase recombination frequency, thus leading to more comprehensive exploitation of sequence space. As a toolbox, RACHITT allows us with more rapid and efficient diversification of genes.

The above DNA shuffling protocols are able to recombine only highly homologous sequences over 70% and produce crossovers in limited regions. However, most enzymes showing similar 3-D structures have less distinct similarity at the level of primary sequences. Various protocols to recombine genes with low sequence homology have been suggested.

In 1999, Benkovic's group developed a novel DNA shuffling tool referred as incremental truncation for the creation of hybrid enzymes (ITCHY) that generates a bunch of recombined gene sequences based on two genes regardless of gene sequence homology (Ostermeier 1999).One of fundamental procedures for ITCHY involves the fragmentation of parent genes with exonuclease with NaCl to limit the digestion speed less than 10 bases/min. The method ITCHY had two shortcomings: recombination must occur between two different parent genes and little hybrid variants were active. With regard to the former, in 2001, Benkovic's group developed a novel tool termed SCRATCHY making the combinatorial approach for target enzymes showing very low homology each other (Lutz 2001). This approach recombines genes by combining gradual truncation for the generation of hybrids DNA. First, gradually cumulative truncation for the generation of hybrids DNA is necessary to generate a complete recombined DNA sequences. This newly created set of DNA is then shuffled again to increase the frequency of crossovers.

In recent years, other novel DNA shuffling tools have emerged, including ADO (assembly of designed oligonucleotides) in 2003, NRR (nonhomologous random recombination) in 2004, NexT (nucleotide exchange and excision technology) in 2005, and RAISE (random insertional deletional strand exchange mutagenesis) in 2006.

Circular Permutation

Circular permutation (CP) of an enzyme starts with the formation of a covalent linkage between the native amino and carboxy terminals by a linker region, followed by random cleavage of the circular peptide to introduce new amino and carboxy terminal elsewhere. The process does not substitute amino acid residues, but only reshuffles the order of residues in the protein. Although modifying enzyme sequences can change the enzyme's folding pathway, few changes occur to the overall structures as indicated by spectroscopic and X-ray crystallographic analyses. Nevertheless, changes can be pronounced near the new terminal and the linker region. For some enzymes, the local changes in structure can turn into substantial conformational changes in the tertiary and quaternary structures, which will have effect on protein stability, catalytic activity or substrate preferences (Yu 2011).CP has offered an exciting new strategy for directed evolution. CP has been employed for improving the catalytic performance of the enzymes such as *Candida antarctica* lipase B(CALB) and xylanase from *Bacillus circulans* (BcX).

Random Insertion/Deletion Mutagenesis

Randomized insertion or deletion of DNA sequences commonly results in the change of length of target enzyme, which will give another chance to make diverse enzymes which cannot be obtained by pointwise mutagenesis. In 1999, Urabe's group described a new method random elongation mutagenesis (REM) (Matsuura 1999). REM adds randomized peptide tails at C-terminal region. In 2000, random insertion and deletion (RID) mutagenesis was suggested to generate the set of mutant enzymes containing nonnatural amino acids (Murakami 2000). Deletion as well as insertion of an arbitrary number of DNA is a key step for RID.

8.2 Rational and Computational Design

8.2.1 Rational Design

Rational designs are dependent on the available information of structures and relationships between sequence and function of enzymes. The fast accumulative number of enzyme structures information deposited on PDB database, and more available homology modeling and molecule substrate docking offer valuable assistance for enzyme engineers to effectively identify active site and locate key residues forming the substrate-binding pocket (Eijsink *et al.* 2004).

Kazlauskas's group studied the mutations to improve enzyme properties such as enantioselectivity, substrate selectivity and catalytic activity. It was revealed that proximal mutations (5–15Å from the active site) are in many cases more effective than distal ones (called as "closer is often better" principle) (Kazlaukas and Bornscher 2009 and Morley 2005). This suggests that focusing mutations near the substrate-binding site instead of the entire enzyme may increase the chance of successful rational design.

Rational design for enzyme engineering involves one or combination of point mutations, recombination of secondary structural elements, and exchange of whole domains or subunits.

If unique amino acid site within an enzyme has been identified as essential one for enzyme function, identification of most suitable amino acid for that position is critically needed. Site saturation mutagenesis (SSM) enables the replacement of unique sites toward the other 19 amino acids at once. However, site-directed mutagenesis (SDM) is used to make specific substitution in the target position. SDM and SSM have become important techniques in the laboratory for enzyme engineering. Whole plasmid PCR *and* overlap *extension PCR* are often used for SDM and SSM, and can also be used to generate the insertion and deletion. Stratagene has developed whole plasmid PCR as the QuikChange® site-directed mutagenesis kit.

It has been suggested that introducing new catalytic activity or changing the substrate specificity of an enzyme would be not achieved by mutating single or few residues. The method grafting loops near active site have extended the use of enzyme engineering. To change a thioester to a β-lactam, insertion, deletion, and substitution of loops as well as amino acid substitutions were made (Park 2006). For the conversion of an esterase from *Pseudomonas fluorescens* into an epoxide hydrolase, the exchange with 20 amino acids at the supposed entrance to active site were needed (Jochens 2009).

Insertion and deletion of loops into enzymes at solvent accessible regions have been interesting points to improve the enzymatic properties or to endow new function. Loops structures have been introduced into enzymes to revise the catalytic properties. Mutant generation by inserting loop structure has been used for O_6-alkylguanine-DNA-alkyltransferase (AGT) catalyzing O_6-propargylguanine as non-native substrate material (Heinis 2010). Deletion mutagenesis has been used to enhance the enzymatic activity and thermostability. Co-improvement of catalytic activity and thermostability of the lipase from *Fervidobacterium changbaicum* has been achieved by deleting lower part of the NC-loop (Li 2012). Noncanonical amino acid incorporation (NCAAI) was suggested as an effective method to support rational design and directed evolution (Zheng 2012).

Semi-rational design applies the knowledge about enzyme sequence, structure and function to choose promising residues with high probability through computational method and generates the knowledge-based library.

For fragment random mutagenesis, only specific fragment of an enzyme, amino acid residue sequence is randomized while the overall scaffold of the enzyme remains unchanged. Multiple sequence alignments can effectively identify functional islands in enzyme sequence. Megaprimer PCR of whole plasmid (MEGAWHOP) can be used for fragment random mutagenesis. In MEGAWHOP, the mutated fragment of gene obtained by epPCR are used as a megaprimer that replaces a homologous region in the template plasmid. After running whole plasmid PCR using the megaprimer, the resultant mixture is treated by *Dpn* I. The *Dpn*I-treated mixture is then transformed into *E. coli* to yield a library.

Different from fragment random mutagenesis, focused directed evolution (or targeted random mutagenesis) focuses on a structural region, a list of key residues forming the reactive site and substrate-binding sites.

The CAST (Combinatorial Active-Site Saturation Test) method (Reetz 2005) is based on analysis of enzyme structure, and during the saturation mutagenesis, several residues located nearby the substrate site are selected and randomized to generate focused candidate sequences. This takes advantages of both saturation mutagenesis at single sites and simultaneous randomization at multiple sites. The CAST tool has been successfully applied in broadening the substrate scope of lipases and enhancing the enantioselectivity of a Baeyer–Villigerase and of an epoxide hydrolase.

Iterative saturation mutagenesis (ISM) (Reetz 2006) performs repeatedly saturation mutagenesis at preselected residues in an enzyme through rational approach. Improvements of enzyme catalytic performances will be the basis in target site selection. When enhancing thermostability, the criterion is based on B-factors. ISM is a knowledge-based approach which needs only small libraries and has proven to be more efficient than all previous systematic efforts.

To enhance the proportion of properly folded chimeras of recombination, Arnold's group developed the SCHEMA, a structure-guided recombination method (Heinzelman 2009). SCHEMA uses the structural data of enzymes to preset borderline of amino acid "blocks," which minimize the average number of side chain. Therefore, SCHEMA could choose the least disruptive crossover locations and design a library of enzyme chimeras. SCHEMA allows recombination of related enzymes with identities as low as 30%, and has been used to improve thermostability of fungal cellobiohydrolases.

Structure-based Combinatorial Protein Engineering (SCOPE) (O'Maille 2002) uses both structural information of enzymes and DNA handling methods simultaneously for the production of a bunch of crossovered gene sequences from genes showing low identity. On the based on exon or domain shuffling concept, SCOPE employs 3-D knowledge to design corresponding gene coding sequences.

8.2.2 Computational Design

Toolbox

Directed evolution to improve enzyme traits faces the bottleneck: for an average-sized enzyme, the sequence space to be explored is huge; nevertheless practical screening capability is limited to thousands or millions of mutants. Virtual screening in silico can analyze a much more numbers of mutants and will increase the success rate to obtain the desired mutants.

ProSAR (protein sequence activity relationships) was presented for in vitro evolution by integrating computational analysis and experimental screening (Fox 2007). ProSAR analysis can be used to classify mutations as beneficial, potentially beneficial, deleterious, or neutral, and these information can be used to design improved mutants in subsequent rounds. ProSAR-driven enzyme engineering reveals that sequence activity relationships can aid directed evolution for novel properties.

Web-based Mutagenesis Assistant Program (MAP) is a toolbox program to help directed evolution strategies by studying the effects of mutational biases of random mutagenesis protocol on any given gene (Table 8.1). The MAP 2.0 server suggests the substitution patterns after specific random mutagenesis. The integrated information can be utilized to choose an experiments direction to elevate the chance of efficient and/or stable mutants in aspect of functionality. Therefore, the MAP web-based program helps preselecction in silico by predicting selection candidates (Table 8.1).

The procedure FamClash analyzes the incompatibilities in engineered hybrids using enzyme family sequence data and ranks the activity of engineered enzymes.

Table 8.1 Websites of the main computational tools for enzyme engineering

Computational tools	Website
MAP	http://map.jacobs-university.de/map3d.html
STAR	http://pprowler.itee.uq.edu.au/star
HotSpot wizard	http://loschmidt.chemi.muni.cz/hotspotwizard/
PoPMuSiC	http://babylone.ulb.ac.be/popmusic
AUTOMUTE	http://proteins.gmu.edu/automute
CUPSAT	http://cupsat.uni-koeln.de/
SPROUTS	http://bioinformatics.eas.asu.medu/sprouts.html
PiSQRD	http://pisqrd.escience-lab.org/

STAR (Site Targeted Amino acid Recombination predictor) generates index showing the probability for disruption in structure triggered by recombination and assists in determining the useful recombination sites.

HotSpot Wizard is a web-based program to identify “hot spots” for changing enzyme traits such as substrate specificity. The “hot spots” are chosen by integrating information sourced from various knowledge.

OptZyme is a newly suggested computational routine to design for activity with a new non-native substrate. Transition state analogs are a starting basis.

PoPMuSiC is a tool for rational computer-aided design of point mutations. It performs all possible point mutations in silico in a given enzyme and the most stabilizing or destabilizing or the neutral mutations with respect to thermodynamic stability are selected based on the solvent accessibility of the mutated position.

CUPSAT is a web program to forecast protein stability differences caused by substitution of residues. This program requires the enzyme structure in PDB format and location of the residues to be mutated.

SPROUTS is a database that can be used for the evaluation of protein stability upon point substitution. And this is a unique resource to analyze the effect of point mutations on protein structures.

PiSQRD is a web program to divide protein into several domains. This program requires the enzyme structure in PDB format.

For commercial program, Discovery Studio is one of the powerful tools for various simulations. First, this program can minimize protein structure to anticipate its structure. CHARMm molecular mechanics simulation program is included in the Discovery Studio, and for quantum mechanics-based studies, DMol progam is also included. Using these two programs, Discovery Studio can efficiently simulate protein structure.

Also, this program can simulate protein–ligand interaction. Binding energy and ligand-bound structure can be calculated and simulated. From the results, effective inhibitor or enzyme with different specificity can be anticipated.

8.2.3 De Novo Enzyme Design

The total design of new biocatalytic enzymes for reactions not catalyzed by native enzymes is a big adventure in enzyme engineering. Recent advances in computational protein design have opened the new era for designing any enzymes for any target reaction. David Baker is a leader in this area. His team developed the Rosetta computational *de novo* enzyme design methodology and obtained the novel enzymes that catalyze the kemp elimination (Röthlisberger 2008), retro-aldol (Jiang 2008), and diels-alder (Siegel 2010) reactions not catalyzed by naturally occurring biocatalysts.*de novo* enzyme design program (Rosetta) can be employed to any targeted reaction in principle. Following four steps are required for *de novo* design of enzyme: (1) selection of a key mechanism for catalysis and model construction mimicking catalytic site, (2) screen the target scaffold to

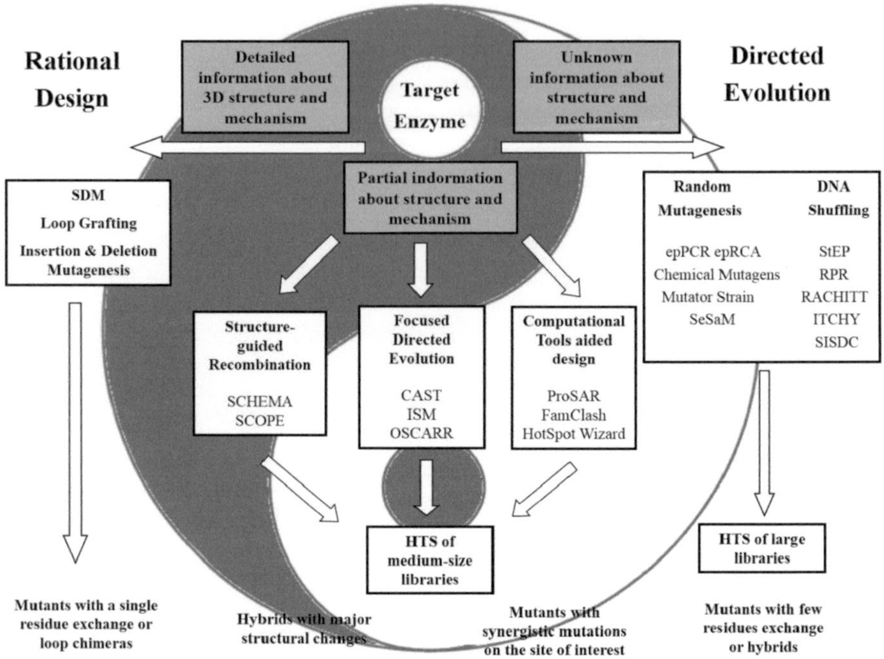

Fig. 8.6 Schematic representation of enzyme engineering tools

realize minimal catalytic sites, (3) optimize the nearby residues to stabilize the transition state, and (4) analysis of resulting design based on ranking.

Enzyme engineering via directed evolution or rational design methods has been a very powerful tool of improving enzyme properties. The tools are fundamentally different by available information. Therefore, the choice of tool is still a case by case depending on the knowledge in depth as well as on practical decision such as available HTS methods as shown in Fig. 8.6. Different methods have their own advantages and drawbacks. Identifying which tool is the best and why particular mutations lead to favored properties will set enzyme engineering forward rapidly. It is unlikely that one method dominates over others, as each problem has different goal, available information, and specifics of the enzyme.

Further Discussion

1. How to decide enzyme's structure? What are the advantages using NMR compared to X-ray crystallography?
2. How to select an appropriate tool of enzyme engineering for a specific enzyme? Summarize the common strategies for directed evolution and rational design. Describe the protocols for site-specific mutation and saturation mutations. Search the cases for enzyme mutations in mammalian cells and plant cells.

3. Solvent accessible surface area (SASA) and B-factor are also important parameters in explaining the enzyme's behavior. How to calculate SASA and B-factor?
4. What is the ultimate goal for enzyme engineering, especially novel enzyme design?
5. What applications can we think on big data or bioinformatics analysis in enzyme engineering area?

References

Bornscheuer UT and Pohl M. Improved biocatalysts by directed evolution and rational protein design. *Current Opinion in Chemical Biology*, 2001, 5:137-143.

Crameri A, Raillard SA, Bermudez E and Stemmer WP. DNA shuffling of a family of genes from diverse species accelerates directed evolution. *Nature*, 1998, 391:288-291.

Eijsink VGH, Bjork A, Gaseidnes S, Sirevag R, Synstad B, Burg B,and Vriend G. Rational engineering of enzyme stability. *J. Biotechnology*, 2004, 113:105-120.

Fox RJ, Davis SC, Mundorff EC, Newman LM, Gavrilovic V, Ma SK, Chung LM, Ching C, Tam S, Muley S, Grate J, Gruber J, Whitman JC, Sheldon RA and Huisman GW. Improving catalytic function by ProSAR-driven enzyme evolution. *Nature Biotechnology*, 2007, 25:338-344.

Greener A, Callahan M and Jerpseth B. An efficient random mutagenesis technique using an *E. coli* mutator strain. *Methods in Molecular Biology*, 1996, 57:375-385.

Heinis C and Johnsson K. Using peptide loop insertion mutagenesis for the evolution of proteins. *Methods in Molecular Biology*, 2010, 634:217-232.

Heinzelman P, Snow CD, Wu I, Nguyen C, Villalobos A, Govindarajan S, Minshull J and Arnold FH. A family of thermostable fungal cellulases created by structure-guided recombination. *Proceedings of the National Academy of Sciences of the United States of America*, 2009, 106:5610-5615.

Jiang L, Althoff EA, Clemente FR, Doyle L, Röthlisberger D, Zanghellini A, Gallaher JL, Betker JL, Tanaka F, Barbas CF 3rd, Hilvert D, Houk KN, Stoddard BL and Baker D. *De novo* computational design of retro-aldol enzymes. *Science*, 2008, 319:1387-1391.

Jochens H, Stiba K, Savile C, Fujii R, Yu JG, Gerassenkov T, Kazlauskas RJ and Bornscheuer UT. Converting an esterase into an epoxide hydrolase. *Angewandte Chemie International Edition*, 2009, 48:3532-3535.

Kazlauskas RJ and Bornscheuer UT. Finding better protein engineering strategies. *Nature Chemical Biology*, 2009, 5:526-529.

Labrou NE. Random Mutagenesis Methods for *in vitro* directed enzyme evolution. *Current Protein and Peptide Science*, 2010, 11:91-100.

Li B, Yang G, Wu L and Feng Y. Role of the NC-loop in catalytic activity and stability in lipase from *Fervidobacterium changbaicum*. *PLoS One*, 2012, 7:e46881.

Lutz S, Ostermeier M, Moore GL, Maranas CD and Benkovic SJ. Creating multiple-crossover DNA libraries independent of sequence identity. *Proceedings of the National Academy of Sciences of the United States of America*, 2001, 98:11248-11253.

Matsuura T, Miyai K, Trakulnaleamsai S, Yomo T, Shima Y, Miki S, Yamamoto K and Urabe I. Evolutionary molecular engineering by random elongation mutagenesis. *Nature Biotechnology*, 1999, 17:58-61.

Morley KL and Kazlauskas RJ. Improving enzyme properties: when are closer mutations better? *Trends in Biotechnology*, 2005, 23:231-237.

Murakami H, Hohsaka T and Sisido M. Random insertion and deletion mutagenesis for construction of protein library containing nonnatural amino acids. *Nucleic Acids Symposium Series*, 2000, 44:69-70.

O'Maille PE, Bakhtina M and Tsai MD. Structure-based combinatorial protein engineering (SCOPE). *Journal of Molecular Biology*, 2002, 321:677-691.

Ostermeier M, Shim JH and Benkovic SJ. A combinatorial approach to hybrid enzymes independent of DNA homology. *Nature Biotechnology*, 1999, 17:1205-1209.

Park HS, Nam SH, Lee JK, Yoon CN, Mannervik B, Benkovic SJ and Kim HS. Design and evolution of new catalytic activity with an existing protein scaffold. *Science*, 2006, 311:535-538.

Reetz MT, Bocola M, Carballeira JD, Zha D and Vogel A. Expanding the range of substrate acceptance of enzymes: combinatorial active-site saturation test. *Angewandte Chemie International Edition*, 2005, 44:4192-4196.

Reetz MT, Carballeira JD and Vogel A. Iterative saturation mutagenesis on the basis of B factors as a strategy for increasing protein thermostability. *Angewandte Chemie International Edition*, 2006, 45:7745-7751.

Röthlisberger D, Khersonsky O, Wollacott AM, Jiang L, DeChancie J, Betker J, Gallaher JL, Althoff EA, Zanghellini A, Dym O, Albeck S, Houk KN, Tawfik DS and Baker D. Kemp elimination catalysts by computational enzyme design. *Nature*, 2008, 453:190-195.

Shao Z, Zhao H, Giver L and Arnold FH. Random-priming in vitro recombination: an effective tool for directed evolution. *Nucleic Acids Research*, 1998, 26:681-683.

Siegel JB, Zanghellini A, Lovick HM, Kiss G, Lambert AR, St Clair JL, Gallaher JL, Hilvert D, Gelb MH, Stoddard BL, Houk KN, Michael FE and Baker D. Computational design of an enzyme catalyst for a stereoselective bimolecular Diels-Alder reaction. *Science*, 2010, 329:309-313.

Stemmer WP. Rapid evolution of a protein *in vitro* by DNA shuffling. *Nature*, 1994a, 370:389-391.

Stemmer WP. DNA shuffling by random fragmentation and reassembly: *in vitro* recombination for molecular evolution. P*roceedings of the National Academy of Sciences of the United States of America*, 1994b, 91:10747-10751.

Wong TS, Tee KL, Hauer B and Schwaneberg U. Sequence saturation mutagenesis (SeSaM): a novel method for directed evolution. *Nucleic Acids Research*, 2004, 32:e26.

Yu Y and Lutz S. Circular permutation: a different way to engineer enzyme structure and function. *Trends Biotechnology*, 2011, 29:18-25.

Zhao H, Giver L, Shao Z, Affholter JA and Arnold FH. Molecular evolution by staggered extension process (StEP) *in vitro* recombination. *Nature Biotechnology*, 1998, 16:258-261.

Zheng S and Kwon I. Manipulation of enzyme properties by noncanonical amino acid incorporation. *Biotechnology Journal*, 2012, 7:47-60.

Chapter 9
Enzyme Catalysis

9.1 Mechanism of Enzyme Activity

9.1.1 Chemical Mechanisms

Enzymes catalyze efficiently various reactions with astounding rates. Understanding the dynamics and molecular mechanism of enzymes has been an important research goal for more than half a century. An enzyme has three-dimensional cleft, called "active site", which is composed of amino acids from different residues. A substrate binds to this active site, through which reaction occurs. Enzymes increase a reaction rate by lowering the activation energy in the formation of a transition state intermediate. The three-dimensional configuration of atoms in the active site is critical for the reaction.

Lock and key model for enzyme action was suggested by Emil Fisher in the end of 19th century. In this model, the lock symbolizes the enzyme and the key is the substrate or substrates. The enzyme molecule provides a uniquely structured template on which the substrate can perfectly attach and interact subsequently (see Fig. 9.1). Once the substrate approaches to the enzyme's active site, an enzyme-substrate (ES) complex is formed. After this transition, products will be released from the enzyme molecule and the enzyme could accommodate another substrate for reaction.

The lock and key model could explain why molecules larger than the enzyme's substrate, but of similar chemical reactivity, are not catalyzed by the enzyme since they cannot access the enzyme's active site. Same analogy can be made for smaller molecules; they might not be catalyzed because they do not fit to the active site to form ES complex. However, Koshland (1958, 1959) found that some smaller molecule which could fit into the active site and has the right chemical stereochemistry did not react. He suggested following model: The substrate causes a change in the three-dimensional structure of the active site geometry; through this the

Y.J. Yoo *et al.*, *Fundamentals of Enzyme Engineering*,
DOI 10.1007/978-94-024-1026-6_9

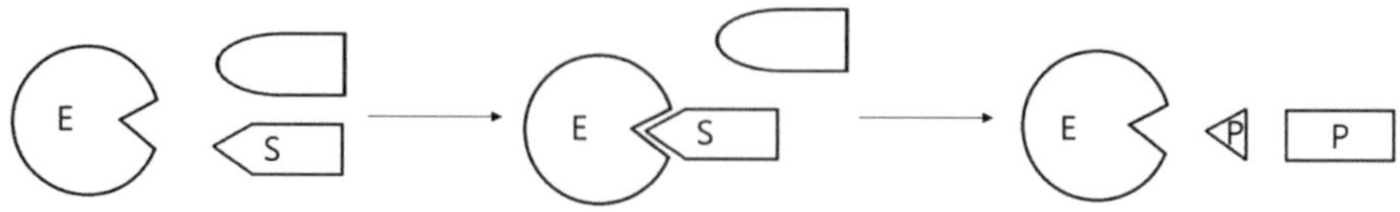

Fig. 9.1 Lock and key model

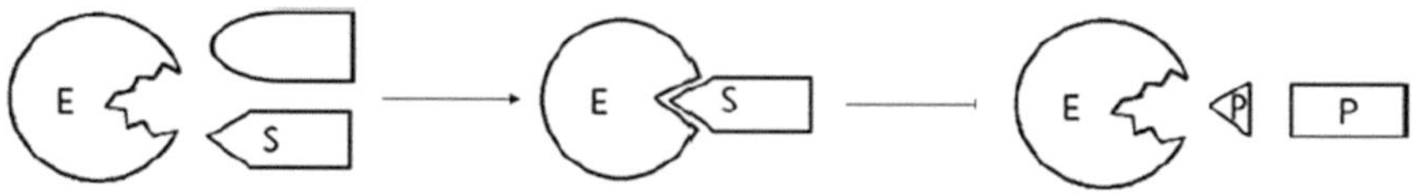

Fig. 9.2 Induced-fit model

substrate can access to the catalytic residues in the right orientation and distance. This theory has been called "induced fit". The induced-fit model shown in Fig. 9.2 assumes that the enzyme is flexible and its active site conformation can be thus changed to fit the substrate.

In induced-fit model, the final orientation of active state—the precise positions of catalytic groups and the residues ready for catalysis—is induced only after the binding of substrate changes the enzyme structure. In contrast, the conformational selection model captures the multiple conformational states adopted by the enzyme, with the substrate having a higher affinity for the active state (Boehr *et al.* 2009). In conformational selection model, the ligand is suggested to bind to the active state of the enzyme. The models presented constitute theories used to rationalize the mechanism by which an enzyme adopts its catalytically functional state (Sullivan and Holyoak 2008). However, these models fail to describe allostericity (change of an enzyme conformation resulting in change in function) and cooperative effects.

Allostric enzymes function by making reversible and noncovalent binding with regulatory compounds called allosteric effectors, which are usually metabolites or cofactors. In many allosteric enzymes, the binding sites for substrate and for the effector are different. Binding of the effector to its specific site causes the conformational change of catalytic subunit, which results in the increased or decreased activity of allosteric enzyme (Fig. 9.3).

9.1.2 Transition State Theory

Transition state occurs as the enzyme-substrate is aligned and transformed into right geometry to give a reaction for product formation. This becomes possible by precise alignment of catalytic groups of enzyme and substrate. Enzymes configure the electronic status by electron transfer, geometric change, protonation and interaction with Lewis acids/bases (Schramm 1998). The summation of individual

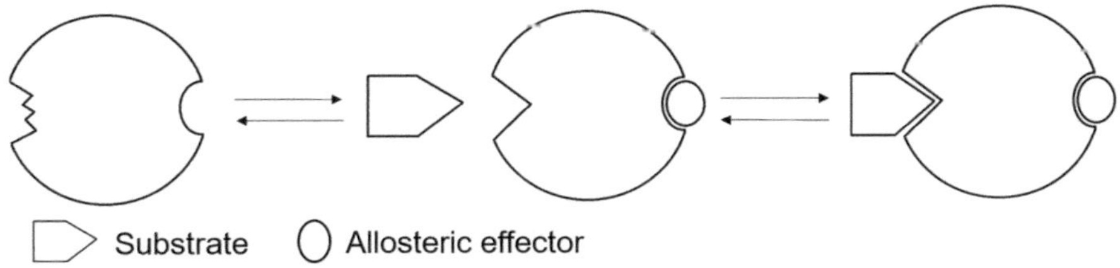

Fig. 9.3 Allosteric enzyme model. After allosteric enzyme interacts with an effector molecule, substrate can thereby bind to the enzyme

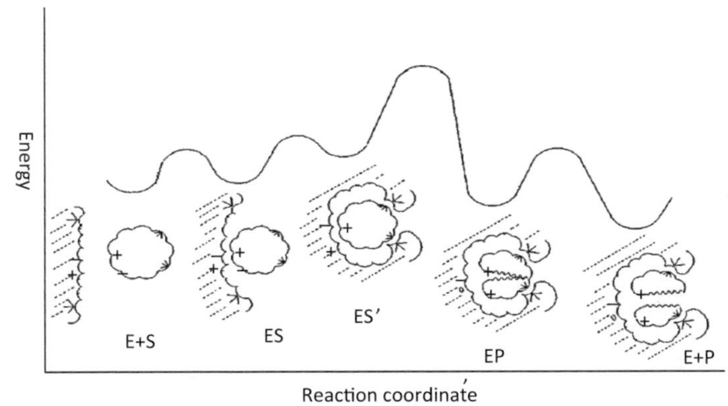

Fig. 9.4 Reaction coordinate diagram for conversion of substrate (S) to enzyme-bound products (EP). Symbols are R–H-bond acceptor, H–H-bond donor, + and − are ionic charges, and > represents hydrophobic sites (Schramm 1998)

weak forces on the substrate results in relocation of electrons for breakage or formation of bonds.

Substrate and enzyme mutually change their structures through realignment and decrease the binding energy of the transition state compared to the enzyme-substrate complex, hence becoming more stable. The hydrogen bonds and ionic bonds between the substrate and the enzyme are dependent on bond angle, interatomic distance, solvent and relative pK_a values. Ionic changes between the enzyme-substrate complex and the transition state are characterized by distinct pK_a values for substrate binding and turnover number, k_{cat} (Parkin and Schramm 1995).

In order for the reaction product to dissociate, the transition state must be relaxed fast. As the bonds are broken to generate a repulsive interaction through electron distributions, the catalytic site is opened and the products are released (see Fig. 9.4).

In structural biology of enzymes, the structure-function paradigm has been progressively reassessed and extended to include dynamic effects as well as some

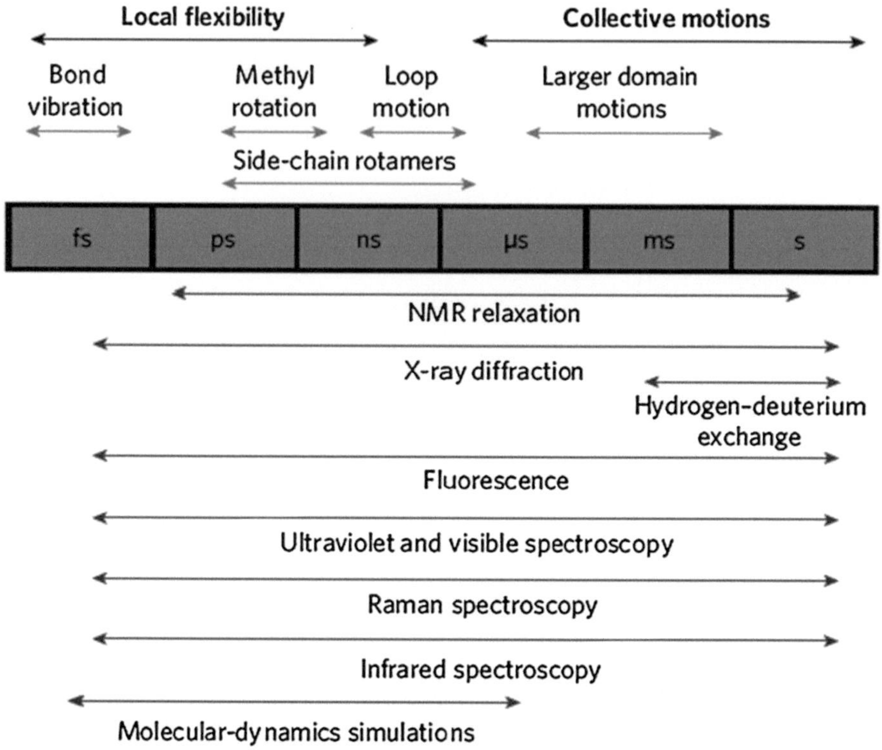

Fig. 9.5 Timescale of protein motions (Henzler-Wildman and Kern 2007)

structural disorder. To understand enzyme action throughly, information on their crystal structures are to be known for every state of enzyme-substrate complex. Since it is impossible to watch experimentally the enzyme action, biophysical methods are used to infer enzyme dynamics from the measured physical properties. Enzymes undergo conformational fluctuations as discussed above, ranging from femto- to milliseconds, and amplitudes (see Fig. 9.5). This conformational range arises from local fast timescale side chain fluctuations and movements of active site loops and slow timescale global structural rearrangements including: secondary structure reorganizations (Skinner *et al.* 2008), interconversion of enzyme fold (Yadid *et al.* 2010), and closed-to-open conformational transitions (Masterson *et al.* 2011) phenomena.

Slow timescales dynamics defines fluctuations between kinetically distinct states that are separated by energy barriers of several kT (the product of the Boltzmann constant and the absolute temperature). Enzyme catalysis, signal transduction and protein-protein interactions occur on the microseconds or slower timescale at physiological temperatures. The individual states of the enzyme conformation at this timescale can be observed real-time or after trapped.

Fast timescale dynamics defines more local, small-amplitude picosecond-to-nanosecond fluctuations at physiological temperature. In contrast to the slow timescale dynamics, fast timescale dynamics requires a statistical description of the distribution (Henzler-Wildman and Kern 2007).

Variety of nuclear magnetic resonance (NMR) spectroscopic methods has been used to document enzyme motions in specific regions. Especially, relaxation dispersion experiments can give both structural and kinetic dynamic information. X-ray diffraction can give information on high resolution structures but a homogeneous crystal is needed. Advances in computational methods provide greater understanding of enzyme dynamics. The exact position of each atom at any given instant can be obtained from one high-resoltuion structure of a protein.

Several explanations for enzyme catalysis mechanism are presented due to the complex enzyme structure and deformation of enzyme and substrate. Induced-fit model proposes that the initial interaction between the enzyme in its native state and the substrate induces a conformational change which leads to stronger binding of the substrate to the enzyme. However, induced-fit model could not explain the real reason why activation energy barrier is lowered during enzyme catalysis. Some enzymes do not also take induced-fit mechanism. Now, the mechanism on how the enzyme lowers the activation energy barrier has been simulated and tested considering the molecular level of the substrate and the enzyme.

Bond strain is derived from the principles of induced-fit mechanism. By conformational change, bonds could be distorted, rotated, polarized or electronically relaxed in substrate or subtle part of the enzyme (Belasco and Knowles 1980 and Clarkson *et al.* 1997). This bond strain can lead reactions to occur easily. However, large bond strain is not expected since enzyme is relatively large and flexible molecule. Bond strain could not explain the transition state stabilization appropriately. It is rather substrate destabilization (Deng *et al.* 1992).

Proximity and orientation can change the substrate and side chain of catalytic residues due to the enzyme structure deformation (Robertus *et al.* 1972). The chemically reacted groups are closed geometrically and the reaction can occur easily. This mechanism is more useful especially for two substrates in one reaction.

Catalytic residues can serve as proton donor and acceptor. They activate nucleophiles and electrophiles or stabilize the leaving group in acid/base reactions. The residues (e.g. Asp, Glu, His, Cys, Tyr and Lys) found near the active site have a role for functional group on acid/base catalysis. Histidine often functions as proton donor or acceptor because its pK_a value is almost neutral. In serine protease, histidine in the active site accepts the proton from serine and serine as nucleophile attacks the peptide bonds of the substrate (Hedsrtom 2002).

Catalytic residues form an ionic bond or covalent bond with the substrate. Ionic bonding is formed by charged amino acids like Asp, Glu, Lys and Arg as well as metal ions. Thermolysin is one of the thermostable peptides which contains zinc ion. This metal ion is indispensable for catalysis because carbonyl oxyanion in the intermediate is stabilized by this zinc ion (Matthews 1988). Ionic interaction or covalent interaction with the substrate or intermediate could stabilize the transition state.

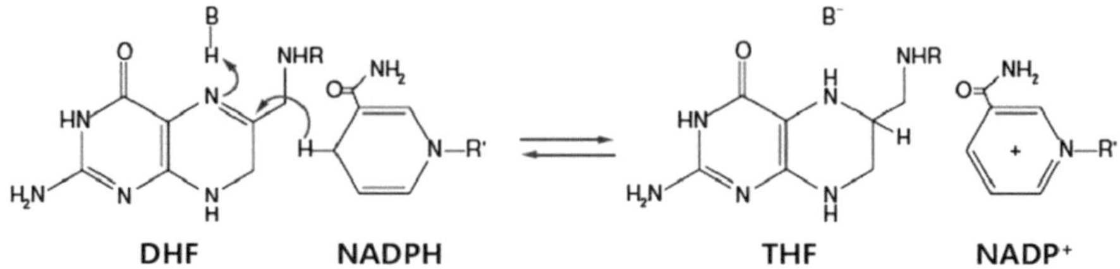

Fig. 9.6 Reaction catalyzed by DHFR

Electrostatic reorganization has been considered as the main reason for lowering the activation energy barrier in computational simulation field. The active site of the enzyme has partial dipoles while the solvent must pay more energy for reorienting its dipoles during the reaction thus enzymatic reaction has lower energetic penalty than the reference reaction in solution (Warshel and Levitt 1976).

Enzyme dynamics is asserted to explain enzyme catalysis mechanism. Enzyme dynamics is broadly accepted as requisite to substrate recruiting, chemical transformation and product release. The fast dynamic motions at femtosecond to picoseconds scale in the transition state can contribute to enzyme reaction. Dynamics couples a promoting vibration to a reactive bond in transition state (Schwartz and Schramm 2009).

As mentioned, MM, QM/MM and QM methods can provide new understanding on the enzyme catalysis. Changes in energy barrier caused by different enzyme or engineered enzyme can be explained by using these methods. Enzyme reactions can be largely classified into 3 steps—binding, reaction, and release. More details on the reaction steps can be also predicted using these methods. Sometimes, product release step is known as the slowest step which means that product release is the rate-limiting step. From this knowledge, engineering of enzyme to accelerate the product release step can be suggested, which remains as future issue.

Case Study: Dihydrofolatereductase (DHFR)

Dihydrofolatereductase catalyzes the reduction of 7,8-dihydrofolate (DHF) to 5,6,7,8-tetrahydrofolate (THF). The cofactor nicotinamide adenine dinucleotide phosphate donates a hydrogen atom to the C_6 atom of the pterinnucleos, with protonation at N_5 as shown in Fig. 9.6 (Nashine *et al.* 2010).

The detailed kinetic scheme by *Escherichia coli* DHFR is shown in Fig. 9.7: the enzyme, cofactor and substrate cycle through five intermediates under cellular conditions. The rate-limiting step is the release of THF from the E·NH·THF complex.

NMR relaxation dispersion was used to characterize the reaction steps. The rate constants obtained by pre-steady state and competitive trapping experiments are consistent with those measured for the transition between the excited and the ground state structures. The hypothesis that the chemical step and the ligand

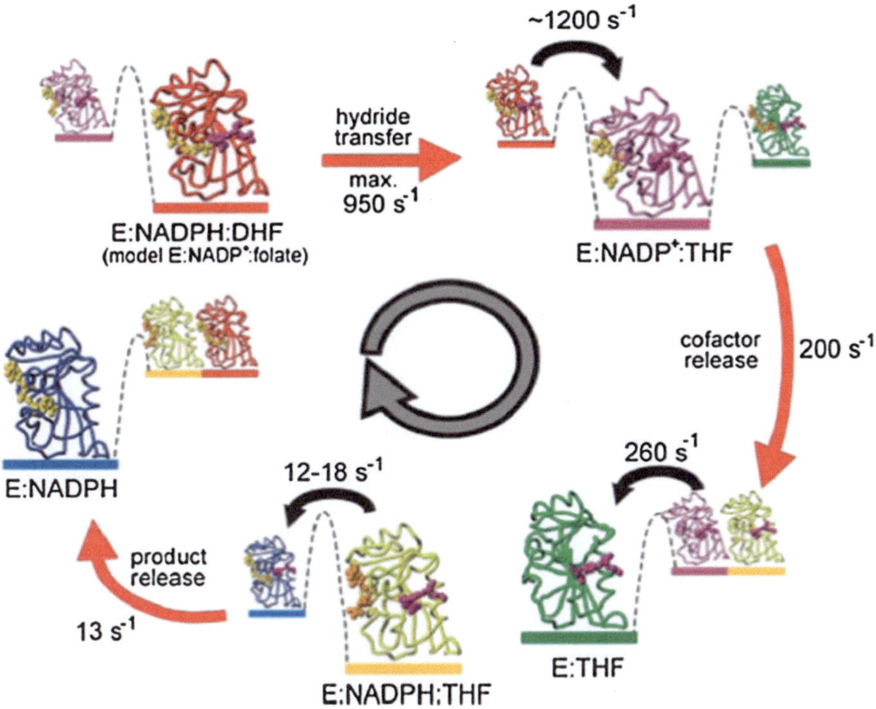

Fig. 9.7 Catalytic cycle of *E. coli* DHFR, showing the conformation of the principal species

release are governed by the rates of conformational changes in the reaction steps is supported by the experiment.

9.2 Enzyme Dynamics and Flexibility

9.2.1 Enzyme Dynamics

Dynamics of enzyme has been associated with catalysis mechanism. Enzyme conformational movements are necessary to enzyme function. The motions by internal dynamics are divided into three: the vibration of individual amino acids including all atoms of the side chains, the movements of amino acid groups like helix and sheet, and domain motion. The time scale of these movements ranges from 10^{-15} to 10^{-1} s for the vibration of atoms, 10^{-9}–1 s for the group movement, and 10^{-7}–10^{4} s for the domain movement. The group movement or rigid body movement may be related to the catalysis mechanism because the time-scale is similar with the k_{cat}.

Each enzyme has different dynamics since the network inside the enzymes are different from each other. Comparing hyperthermophilic and mesophilic homologs of adenylate kinase showed the difference in residual flexibility in lid structure as measured by nuclear magnetic resonance (NMR) relaxation which mobility is the rate-limiting step in the reaction (Wolf-Watz *et al.* 2004). The hyperthermophilic homolog showed that catalytic activity at ambient temperature and the lid-opening rate decreased. High flexibility explains high catalytic activity of psychrophilic enzymes at low temperature.

Debye-Waller factor, also termed as B-factor, is one of the variables representing enzyme flexibility. It is calculated in X-ray crystallography data, having a unit Å^2. B-factor describes the vibration of atoms. Atoms with high B-factor is usually in disordered or flexible structure, while atoms with low B-factor are in ordered and rigid structure. The spin relaxation in NMR has also been used to access the dynamics and mobility of enzymes. NMR relaxation could check fast ps and ns motions as well as slower microsecond to millisecond motions (Bracken 2001). The spin removal of some residues within the protein represents that the residues move so as to the fast indirection of spin means more mobile and flexible atom.

Root mean square deviation or distance (RMSD) and root mean square fluctuation (RMSF) are commonly used to determine enzyme flexibility in computational simulations. During dynamics simulation, most atoms comprising the enzyme move. Two enzyme structures are superimposed then the distance from the original position and the final one is measured to obtain the RMSD value. RMSF is averaged value of RMSD taken over a certain period of time to represent a more reliable comparison for residual flexibility.

Figure 9.8 shows one example of recent understanding on the enzyme's structure. Compared to classical understanding which considers enzyme as one big globule, current understanding considers enzyme as mechanical machine functioning chemical reaction. In this machine, some domain mainly needs flexibility for catalytic motion, the other domains require rigidity to support the catalytic doman. Further analysis on enzyme's mechanical structure having catalytic residues can be used to understand the enzyme structure *versus* function and eventually design a novel enzyme in the future.

Case Study: Flexibility Analysis

Many mutants were reported on the bacteriophage T4 lysozyme, where activity-enhaced mutants were included. Flexibility analysis was performed for the activity-enhanced mutants of bacteriophase T4 lysozyme. Mutations of the amino acid residues on the edge of the helices towards more flexible amino acids showed the activity-enhanced results. This tendency was also confirmed using B-factor analysis (Hong *et al.* 2014). Flexibility change can be used as a tool in enhancing the enzyme activity and was successfully applied for the increase of lipase enzyme (Hong and Yoo 2013).

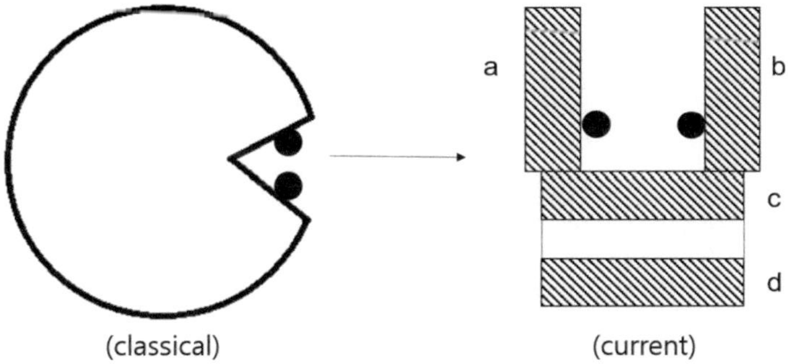

Fig. 9.8 Schematic diagram on the understanding of enzymes. a, b: active site domain containing catalytic residue, c, d: active site-supporting domain • : catalytic residue. Each domain mainly consists of α-helix with/without β-sheet

Further Discussion

1. What is the source of the energy for enzyme motion?
2. Enzyme reaction occurs through many steps, such as attachment of substrate to enzyme, reaction at catalytic site, and detachment of product(s) from enzyme's pocket. How can we find the bottleneck in enzyme's motion and make debottlenecking?
3. Discuss the relationship between flexibilities of amino acids and flexibility of enzymes. Can we change the flexibility of the enzyme by changing the amino acids at specific location, which eventually change the activity of the enzyme?
4. Enzymes show conformational motions and thus can be considered to move like a elastic body. However, it is not easy to understand the details of the conformational motion, instead domain-based analysis can be performed to understand the enzyme's motion. Discuss how domain can be defined and the domain-based analysis can be used.

References

Belasco JG and Knowles JR. Direct observation of substrate distortion by triosephosphate isomerase using Fourier transform infrared spectroscopy. *Biochemistry*, 1980, 19:472-477.

Boehr DD, Nussinov R, and Wright PE. The role of dynamic conformational ensembles in biomolecular recognition. *Nature Chemical Biology*, 2009, 5(11):789-796.

Bracken C. NMR spin relaxation methods for characterization of disorder and folding proteins. *Journal of Molecular Graphics and Modelling*, 2001, 19:3-12.

Clarkson J, Tonge PJ, Taylor KL, Dunaway-Mariano D and Carey PR. Raman study of the polarizing forces promoting catalysis in 4-chlorobenzoate–CoA dehalogenase. *Biochemistry*, 1997, 36:10192–10199.

Deng H, Zheng J, Sloan D, Burgner J and Callender R. A vibrational analysis of the catalytically important C4–H bonds of NADH bound to lactate or malate dehydrogenase: ground-state effects. *Biochemistry*, 1992, 31:5085–5092.

Hedstrom L. Serine protease mechanism and specificity.*Chemical Reviews*, 2002, 102:4501-4523.

Henzler-Wildman K and Kern D. Dynamic personalities of proteins. Nature, 2007, 450:964-972.

Hong SY and Yoo YJ. Activity enhancement of *Candida antarctica* lipase B by flexibility modulation in helix region surrounding the active site. *Appl. Biochem. Biotechnol.*, 2013, 170: 925-933.

Hong SY, Park HJ and Yoo YJ. Flexibility analysis of activity-enhanced mutants of bacteriophage T4 lysozyme.*J. Mol. Catalysis B : Enzymatic,* 2014, 106:95-99.

Koshland DE. Application of a theory of enzyme specificity to protein synthesis. *Proc Natl Acad Sci USA*, 1958, 44:98-104.

Koshland DEJ. Mechanism of transfer enzyme in: *The Enzymes, revised edition* (Boyer P, Lardy H, and Myrback K ed.). Academic Press, New York, 1959, 305-346.

Masterson LR, Shi L, Metcalfe E, Gao J, Taylor SS and Veglia G. Dynamically committed, uncommitted, and quenched states encoded in protein kinase A revealed by NMR spectroscopy. *Proceedings of the National Academy of Sciences*, 2011, 108:6969–6974.

Matthews BW. Structural basis of the action of thermolysin and related zinc peptidases. *Accounts of Chemical Research*, 1988, 21:333-340.

Nashine VC, Hammes-Schiffer S and Benkovic SJ. Coupled motions in enzyme catalysis. *Current Opinion in Chemical Biology*, 2010, 14:644–651.

Parkin DW and Schramm VL. Binding modes for substrate and a proposed transition-state analogue of protozoan nucleoside hydrolase. *Biochemisty*, 1995, 34:13961-13966.

Robertus JD, Kraut J, Alden RA and Birktoft JJ. Subtilisin; a stereochemical mechanism involving transition-state stabilization. *Biochemistry*, 1972, 11:4293–4303.

Schramm VL. Enzymatic transition states and transition state analog design. *Annual Review of Biochemistry*, 1998, 67:693-720.

Schwartz SD and Schramm VL. Enzymatic transition states and dynamic motion in barrier crossing. *Nature Chemical Biology*, 2009, 5:551-558.

Skinner JJ, Wood S, Shorter J, Englander SW and Black BE. The Mad2 partial unfolding model: regulating mitosis through Mad2 conformational switching. *Journal of Cell Biology*, 2008, 183:761–768.

Sullivan SM and Holyoak T. Enzymes with lid-gated active sites must operate by an induced fit mechanism instead of conformational selection. *Proceedings of the National Academy of Sciences*, 2008, 105(37):13829-13834.

Warshel A and Levitt M. Theoretical studies of enzymatic reactions: dielectric, electrostatic and steric stabilization of the carbonium ion in the reaction of lysozyme. *Journal of Molecular Biology*, 1976, 103:227-249.

Wolf-Wats M, Thai V, Henzler-Wildman K, Hadjipavlou G, Eisenmesser EZ and Kern D. Linkage between dynamics and catalysis in a thermophilic-mesophilic enzyme pair. *Nature Structural & Molecular Biology*, 2004, 11:945-949.

Yadid I, Kirshenbaum N, Sharon M, Dym O and Tawfik DS. Metamorphic proteins mediate evolutionary transitions of structure. *Proceedings of the National Academy of Sciences*, 2010, 107:7287–7292.

Chapter 10
Specificity of Enzymes

10.1 Substrate Specificity

10.1.1 Understanding of Substrate Specificity

The enzymatic reaction starts with the binding of the substrate to the enzyme. When the substrate approaches to the active site of enzyme, the electrostatic microenvironment in the substrate-binding region changes to make the reaction proceeds to form the final products. In this process, the correct and exact binding of substrate to the substrate-binding site of enzyme is important. Thus, the specificity of enzyme is closely related to the binding configuration and affinity of the substrate to the active site of enzyme. One example of the substrate specificity is protease. There are many types of proteases, such as trypsin, chymotrypsin, and thrombin, which cleave peptide bond but acting on different peptide bonds. For example, trypsin acts on lysine or argine, chymotrypsin on tyrosine, tryptophan, or phenylalanine, and thrombin on arginine-glycine peptide bond.

As explained earlier, classical models which explain the enzyme-substrate reaction were based on the complementarity between the geometrical shapes of substrate and enzyme active site. The substrate-binding region of enzyme has a particular geometric shape that is complementary to that of the substrate molecule. This indicates that enzymes react with specific compounds of geometrically similar structure. The specificity of an enzyme with a substrate can be explained by "Lock and key" model. In this model, the lock and key correspond to the enzyme and the substrate, respectively, and only the correctly shaped key can fit into the key hole (active site). This theory is based on the "rigid enzyme" model focused on the geometric complementarity between the shape of an enzyme and a substrate.

Y.J. Yoo *et al.*, *Fundamentals of Enzyme Engineering*,
DOI 10.1007/978-94-024-1026-6_10

The rigid enzyme model assumed in the lock and key model is not always valid for the explanation of experimental phenomena. A modified model called "Induced-fit theory" was postulated on the assumption that enzyme is partially flexible. The structure of enzyme is slightly changed by the initial binding of substrate to enzyme, and then the resulting shape of the enzyme is determined to accept the substrate with exactly right configuration. Some compounds can bind to the enzyme without undergoing a reaction. The phenomenon cannot be explained by the Lock and Key model but by the induced-fit theory; it is because even though the substrate can bind to the enzyme, the changed enzyme structure is not proper for the corresponding reaction to be occurred. Only the specific substrate can induce the proper configuration of the substrate-binding site.

How to predict the substrate specificity? Preliminary *in silico* high-throughput method was suggested (Tyagi and Pleiss 2006). By docking substrate into the enzyme, transition state analogous intermediates can be obtained. By examining the effect of side chain orientation and the affinity of the enzyme-substrate complex, substrate specificity of the enzyme can be predicted and can be used for further applications such as virtual screening of potential substrates and enzyme engineering.

10.1.2 Engineering Substrate Specificity

There has been growing interests in the use of enzymes for the production of industrial chemicals, biofuels, and high-value pharmaceuticals. Enzymes have advantages in the synthetic processes for their safe, energy saving, and environment-friendly natures. However, natural enzymes can be limited by their finite catalytic repertoire in the application of broad areas of biotechnology. It is because the natural enzymes evolved for the needs of their natural role, not for the human benefit. Thus, they are frequently not suitable for many industrial applications, and do not satisfy the requirements of industrial biotechnology. For instance, the development and improvement of the biosynthetic pathways for metabolic engineering only with the natural enzymes could be restricted by the limited catalytic diversity. Therefore, understanding and engineering of the enzyme-substrate specificity is one of the most important issues in the area of enzyme applications. In the theory of enzyme-substrate specificity, the "binding" of substrate to the active site of enzyme is the main factor that determines the specificity of an enzyme In this regard, the engineering of the active site structure of enzyme is the essential design approach for the engineering of enzyme-substrate specificity; and thus, the rational and computational approach is mostly aimed at the modification of the substrate-binding pocket (Wijma and Janssen 2013). The key steps for the engineering are (1) selection of the mutation sites in the binding pocket, and then (2) mutation of the selected residues to appropriate counterparts.

Among the nonstructure-based approaches in protein engineering, directed evolution is a representative of random approach that requires no knowledge of the enzyme structure or mechanism (Arnold and Volkov 1999). Directed evolution can be achieved either by DNA shuffling or error-prone PCR. This method requires no structural information with practically effective results. However, it usually needs multiple rounds of evolution and huge amount of variants to be screened. It is a time- and labor-consuming process. However, directed evolution is still the most widely used method due to the practical effectiveness of this approach.

The random approaches do not consider the relationships between the structure and function of enzymes, so that it is hard to understand the fundamental role of the applied mutations in the catalysis of the enzyme. Because of the large increases in the accumulation of enzyme crystal data and knowledge of enzyme catalysis, rational and computational approach or semi-rational design has become more favorable for the engineering of enzyme-substrate specificity. In addition, this approach provides comprehensible information to understand the structure–function relationships involved in enzyme specificity.

In rational design, based on the data of protein structures, one can propose mutations which can be introduced by site-directed mutagenesis. Rational design is advantageous in that it can increase the probability of effective mutations and reduce the library size, and finally save effort and time for the library screening. Moreover, this method requires no high-throughput assay system that can significantly reduce the experimental cost. The common process for rational design is performed by reshaping the structure of substrate-binding pocket considering the geometric and/or physico-chemical complementarity of the enzyme with the target substrate. In rational design, the commonly used strategies for the redesign procedure are (1) to increase the binding pocket space to allow for a bigger substrate, (2) to decrease the binding pocket space to allow for a smaller substrate, and (3) to control the substrate-binding mode by adjusting the molecular interactions between the substrate and the amino acid residues in the active site (Manna and Mazumdar 2010; Mouratou *et al.* 1999; Sinclair *et al.* 1998). With the strategies, the mutation sites are selected and the amino acids to be substituted are decided. However, even though the approaches are based on the structural information of the enzyme and substrate, this is a qualitative approach that largely depends on the insight and ability of the researchers. In addition, the strategies are focused mainly on the "binding" character of the enzymes.

Recently, computational design is getting interest as an advanced engineering tool for the improvement of enzymes (Kiss *et al.* 2013). In this approach, the selection of mutation sites and decision of the substituents can be automatically processed according to the applied computational algorithms. The calculations are performed for anything ranging from single to dozens of mutations at a time (Gordon *et al.* 2012; Khare *et al.* 2012; Murphy *et al.* 2009). Computational enzyme design algorithms seek to identify appropriate protein structures that are low in energy when folded to the prespecified target. Among the computational design programs, the most widely used are Rosetta design and K* algorithm. After putting the substrate compound in the active site of the target enzyme, in Rosetta design, amino acid residues surrounding the substrate are mutated and optimized to ensure good packing and fold stability (Kiss *et al.* 2013). On the other hand,

K* is an ensemble-based enzyme redesign algorithm that considers all rotamer-based conformations for a protein-substrate complex, including all combinations of the rotamers for the active site residues as well as the substrate. However, this approach still has some problems with the large amount of computational load. Thus, the computing time and efficiency depend largely on the power of the computer system. The accurate modeling with the approach is still a challenge due to the limited accuracy of the applied algorithms (Tyka *et al.* 2012).

A semi-rational approach that combines the rational and random approach also can be applied to improve the function of enzymes. In this case, the application of saturation mutagenesis on the active site of enzyme is the most widely used approach for the design of enzyme-substrate specificity (Andrews and McLeish 2013; Gao *et al.* 2013; Xie *et al.* 2014). Because the whole random approach can generate too many libraries that have the mutation sites far from the active site, focused mutagenesis on the active site residues can improve the efficiency of the design approach. The semi-rational approach can generate synergistic effects by taking advantages of the rational and random approach. Therefore, it is a practically efficient approach for the design of improved enzyme variants. However, this approach still depends on the experimental efforts in the mutation and selection step of the approach. The experimental dependency is significantly increased if the mutation sites are extended when considering all the amino acids as the substituents.

Case Study: Engineering Substrate Specificity of 3-Hydroxybutyrate Dehydrogenase (Yeon *et al.* 2013)

The substrate specificity of 3-hydroxybutyrate dehydrogenase (3HBDH, EC 1.1.1.30) from *Alcaligenes faecalis* was changed to convert levulinic acid to the 4-hydroxyvaleric acid by rational design (Fig. 10.1). Molecular docking simulation and enzyme-substrate interatomic contact analysis was applied for the engineering of 3HBDH. The binding energy and donor–acceptor distances were employed as key design parameters considering the H-transfer mechanism of the catalysis.

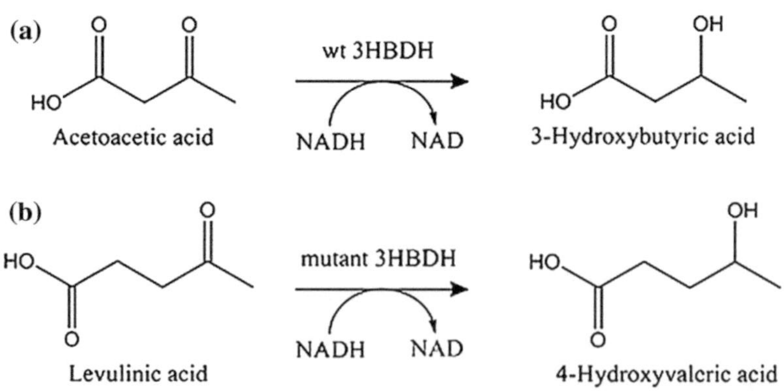

Fig. 10.1 **a** The reduction of acetoacetic acid by wild-type 3HBDH, **b** The reduction of levulinic acid by mutant 3HBDH

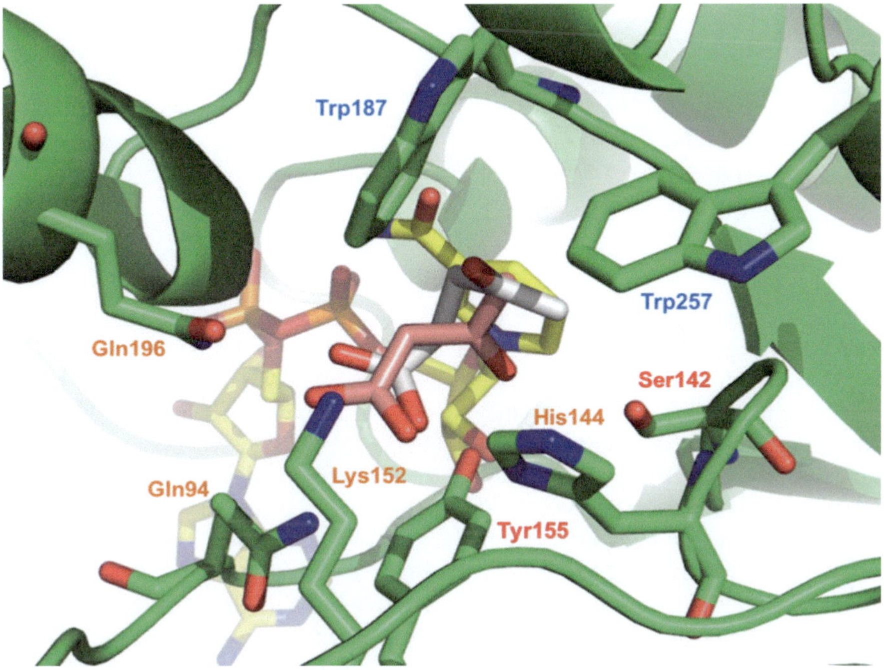

Fig. 10.2 Substrate contacting residues of 3HBDH; substrates (*pink* acetoacetate, *white* levulinic acid), contacting residues (*orange* in the entrance region of binding pocket, *red* catalytic residues, *blue* in the inside region of binding pocket)

Through molecular docking and enzyme-substrate interatomic contact analysis (Sobolev *et al.* 1999), six residues located in the substrate-binding site, Gln94, His144, Lys152, Trp187, Gln196, and Trp257, were selected as initial target residues for mutation. By rational design approach, the six residues were mutated to Gln94Asn, His144Leu, Lys152Ala, Trp187Phe, Gln196Asn, and Trp257Phe. By measuring the specific activities of the six mutants with levulinic acid, His144Leu mutant which has an activity was selected as the mutant for further improvement (Fig. 10.2).

From the first design results, His144 was identified as a key residue for the engineering of 3HBDH. Additional single and combitatorial mutations were tested. Based on this result, His144Leu/Trp187Phe was selected as the best double mutant, in this study, for the enzymatic reduction of levulinic acid. The kinetic parameters (K_m and k_{cat}) for the mutants including the wild-type 3HBDH, are shown in Table 10.1. The H144L mutant showed an 8.6-fold increased catalytic activity (k_{cat}/K_m) relative to the wild-type and the H144L/W187F double mutant showed a 33.4-fold increase.

Case Study: Change of NAD(P) Specificity (Watanabe *et al.* 2005)

Xylose is one of the major compounds of hemicellulose. *Saccharomyces cerevisiae*, widely used for ethanol production, cannot uptake xylose as a carbon source,

Table 10.1 The structural and kinetic parameters of 3HBDHs with levulinic acid

3HBDH	ΔG_b (kcal·mol^{-1})	d_1 (Å)	d_2 (Å)	K_m (mM)	k_{cat} (min^{-1})	k_{cat}/K_m (M^{-1} s^{-1})
WT	−10.3	4.4	6.7	0.78 ± 0.08	0.36 ± 0.05	7.68 ± 1.12
H144L	−9.6	4.0	5.0	0.91 ± 0.09	3.61 ± 0.48	65.94 ± 8.42
H144L/W187F	−11.0	3.3	3.4	0.64 ± 0.07	9.83 ± 1.08	256.54 ± 30.34

and must be engineered to utilize glucose and xylose at the same time. For this purpose, the genes involved in xylose metabolism were introduced from other organisms such as *Pichia stipitis*. However, intercellular redox imbalance occurs because of the different coenzyme specificity of xylose reductase (NADPH-dependent) and xylitol dehydrogenase (NAD^+-dependent). To improve the productivity, a change of specificity of xylitol dehydrogenase (XDH) to $NADP^+$ is needed. In this study, XDH from *P. stipitis* (PsXDH) was used as a target enzyme.

XDH belongs to dehydrogenase/reductase (MDR), which constitutes a large enzyme superfamily. XDH belongs to the polyol dehydrogenase (PDH) subfamily that includes sorbitol dehydrogenase (SDH) and other enzymes. In comparison with $NADP^+$-dependent SDH from *Bemisia argentifolii*, Asp207, Ile208, Phe209, Asn211 in PsXDH were chosen as mutation targets to change the coenzyme specificity. The phosphate ion in $NADP^+$ is found adjacent to Ala199, Arg200, and Arg204 in SDH. Ala199-Arg200 in SDH are homologous to Asp207-Ile208 in PsXDH. Neighboring residues were also chosen for expression of single, double, triple, and quadruple mutants (Fig. 10.3).

Triple mutant ARS (D207A, I208R, F209S) and quadruple mutant (ARS + N211R) showed improved $NADP^+$ dependent activity. However, thermostability was decreased. To improve thermostability, zinc metal ion was added to the enzyme. Many MDRs have one zinc atom at the catalytic site for activity. Some possesses structural zinc atom, which role has not yet been clarified, although the role seems to maintain the quaternary structure. The zinc atom was coordinated with four cysteine residues in the *B. argentifolii* SDH. From structural comparison, PsXDH was mutated to S96C, S99C and Y102C to introduce additional zinc atom. With triple/quadruple mutant, zinc atom introduced PsXDH shows high thermostability and $NADP^+$ dependent activity.

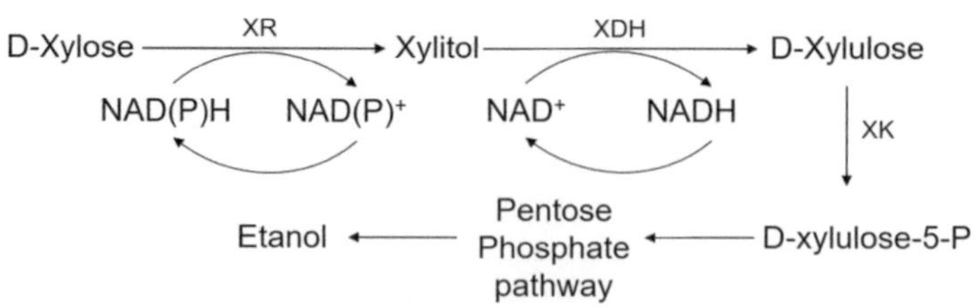

Fig. 10.3 Xylose reaction pathway for the production of ethanol

10.2 Enantioselectivity of Enzymes

Growing need for optically pure compounds in flavor chemical, agricultural, and pharmaceutical industries has driven the development in enantioselective synthesis (Rouhi 2004). Along with the interest in environmentally benign and sustainable processes, enzymes have drawn attention as intrinsically chiral and efficient catalysts (Otten *et al.* 2009). Configurational complementarity between the enzyme's active site and the substrate enantiomers is the single most important factor to determine which enantiomer reacts faster than the other, because the enantiomeric pair shares the same chemical composition and bonds but only differs by 3-dimensional configuration. An enzyme tends to have an asymmetrical active site which is tailor-made for the productive binding of the fast-reacting enantiomer. The slow-reacting enantiomer has limited interaction with the active site or binds in a nonproductive binding mode where the catalysis cannot take place properly. On the other hands, unnatural substrates do not have optimized configurations for the active site, leading to relatively low enantioselectivities. Most enzyme engineering strategies to control enantioselectivity have evolved to affect the configurational complementarity by optimizing the interaction between the active site and the desired enantiomer while disrupting the interaction with the undesirable enantiomer.

10.2.1 Measurement of Enantioselectivity

Enantioselectivity can be measured in two different ways. One is enantiomeric excess, ee, which is defined as purity of one enantiomer in a sample. The equation is as follows:

$$\text{ee} = \frac{F_{\text{major}} - F_{\text{minor}}}{F_{\text{major}} + F_{\text{minor}}} \tag{10.1}$$

F_{major} is the fraction of major enantiomer and F_{minor} is the fraction of minor enantiomer in a mixture. For example, in a mixture of enantiomers with 99:1 ratio, the calculated ee is 98%.

Another way to measure enantioselectivity is enantiomeric ratio, E, which is the selectivity of a catalyst or reagent for one enantiomer over the other and calculated as follows:

$$\begin{aligned} E &= \frac{\text{rate of formation of fast enantiomer}}{\text{rate of formation of slow enantiomer}} \\ &= \frac{K_{\text{cat}}}{K_m}(\text{fast}) / \frac{K_{\text{cat}}}{K_m}(\text{slow}) \end{aligned} \tag{10.2}$$

10.2.2 Kinetic Resolution

In kinetic resolution, the reaction rate of two enantiomers of a substrate is different in an enzyme-mediated reaction (Fig. 10.4) (Keith *et al.* 2001).

In Fig. 10.4, the R-enantiomer is favored due to lower activation energy than that of the S-enantiomer, and the resolution capability is higher with greater $\Delta\Delta G‡$. After the conversion of each substrate enantiomer into its corresponding product enantiomer, the faster reacting enantiomeric product is separated from the slower reacting enantiomeric substrate by conventional separation procedures, which is more efficient than attempting to separate an enantiomeric pair of the same compound by chiral separation methods.

The enantiomeric excess is affected by the conversion rate, where the $ee_{substrate}$ increases along with the increasing conversion but $ee_{product}$ is decreased at the same time. As the fast-reacting enantiomer is converted to the product at a higher rate than the slow-reacting one, an excess of slow-reacting substrate enantiomer is recovered with higher purity for a larger conversion.

10.2.3 Dynamic Kinetic Resolution

Kinetic resolution has limitations that it requires separation of the product from remaining substrate, and there is a drop in enantiomeric purity as the process nears 50% conversion. The theoretical yield is 50% at maximum, because only one enantiomer of the racemate is utilized. In order to overcome these issues, a

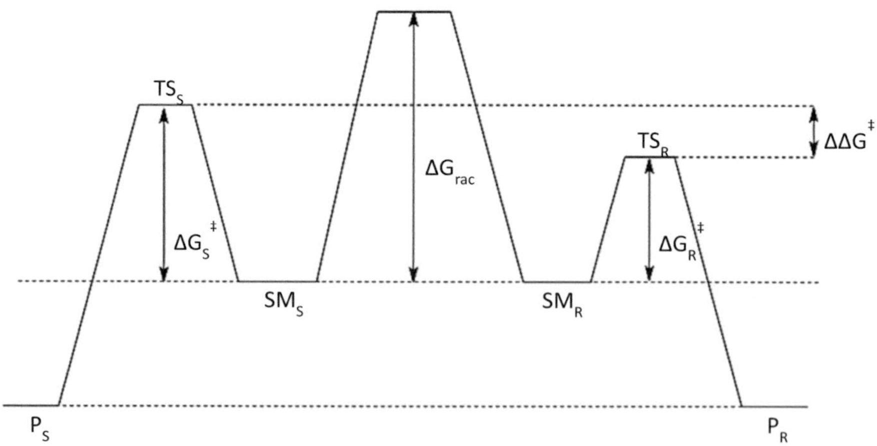

Fig. 10.4 Free energy diagram for kinetic resolution. SM_S and SM_R are the substrate enantiomers; P_S and P_R are the product enantiomers; TS_S and TS_R are the transition states, with $\Delta G_S‡$ and $\Delta G_R‡$ as activation energy, respectively; $\Delta\Delta G‡$ is the difference in the activation energies of the two enantiomers; ΔG_{rac} is the activation energy of racemization between the enantiomers

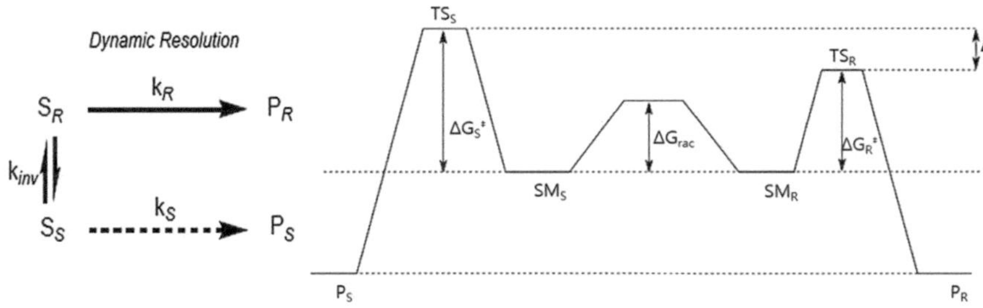

Fig. 10.5 Reaction scheme for dynamic kinetic resolution and free energy diagram for dynamic kinetic resolution. ΔG_{rac} is lowered to a level below the $\Delta G_S‡$ and $\Delta G_R‡$ by a racemase enzyme

notion of dynamic kinetic resolution can be coined. In dynamic kinetic resolution, a racemase is utilized to lower the ΔG_{rac}, allowing interconversion between the substrate enantiomers (Fig. 10.5).

Racemization and kinetic resolution steps can be used simultaneously to increase the theoretical maximum yield from 50 to 100% for a single enantiomer. This assumes a Curtin–Hammett system in which the composition of products is determined by the relative free energies of the transition state, rather than the equilibrium distribution of the starting material (Carey and Sundberg 1984).

For an optimized dynamic kinetic resolution, the kinetic resolution step should be irreversible and the enantiomeric ratio ($E = k_R/k_S$) should be large to ensure high enantioselectivity (Strauss *et al.* 1999). Furthermore, the racemization rate (k_{inv}) should be at least equal or greater than the reaction rate of the fast enantiomer (k_R) to avoid depletion of the S_R. In case the E value is only moderate, k_{inv} should be greater than k_R by a factor of ~10. Any spontaneous racemization of the product should not occur.

10.2.4 Deracemization and Stereoinversion

In cases of compounds with a configurationally stable chiral center, such as secondary alcohols, it is difficult to achieve in situ racemization. Deracemization is used in these cases to convert the chiral center to achiral, for example by oxidizing the alcohols to ketones. This changes the tetrahedral configuration into a trigonal planar, removing the chirality (Fig. 10.6). One enantiomer of a racemic alcohol is oxidized by enantioselective dehydrogenase while the other remains unreacted. The ketone is then reduced again by a different enzyme with opposite enantioselective preference to create an inverted chiral center.

For optimal biocatalytic deracemization and stereoinversion, at least one of the two redox reaction steps has to be irreversible to pay for the entropy balance. Since the overall process constitutes of an oxidation and a reduction reaction, the net redox balance is zero. And no external cofactor recycling is necessary in an ideal case (Strauss *et al.* 1999).

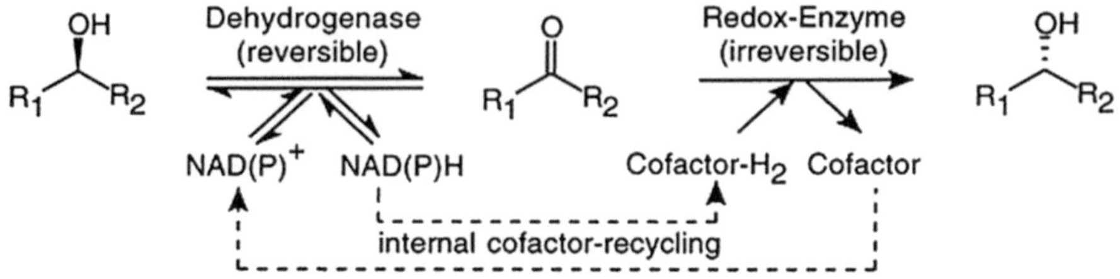

Fig. 10.6 Scheme for deracemization and stereoinversion of a secondary alcohol

10.2.5 Desymmetrization

Selectivity is also the bases of this approach. Starting with symmetric substrate like meso-compounds or prochiral precursors, an enzyme converts only one of the functional groups, which results in desired molecule in theoretically 100% yield. Even though additional recycling or combination steps are required in kinetic resolutions or deracemizations, desymmetrization can be performed in one step.

10.2.6 Asymmetric Synthesis

In asymmetric synthesis, the substrates are prochiral precursors which can be turned into chirally active compound through enantio-selective reaction. The approach offers enhancement in product quantity and enantiomeric ratio. Table 10.2 shows the feasibility of producing fine chemical and pharmaceutical products on commercial scale.

Enzyme's enantioselectivity depends also on the geometry of the enzyme's structure, especially active site pocket. As an example, for subtilisin-catalyzed transesterification reaction among vinyl butyrate and S-, R-enantiomers of chiral secondary alcohols, higher reactivity of S-enantiomer can be explained using steric hinderance and tight pocket size in the active site of the enzyme (Fitzpatrick *et al.* 1992).

Chirally pure D and L amino acids have great demands as versatile building blocks of peptide as well as key constituents of inject solutions, food ingredients and live-stock feed ingredients. One of the best established processes uses acylase to produce L-amino acids from the racemates of N-acetyl-DL-amino acids. Acylase specifically catalyzes the cleavage of N-acetyl-L-amino acid, which in turn produces L-amino acid and leaves N-acetyl-D-amino acids. Through ion-exchange or crystallization L-amino acid was purified as product while N-Ac-D-amino acids are recycled by thermal racemization. Of course, D-amino acid can be obtained if D-specific aminoacylase is applied. In situ racemization using racemase can yield significant improvement by eliminating cost intensive racemization steps. German company Degussa employed acylase enzyme originated from

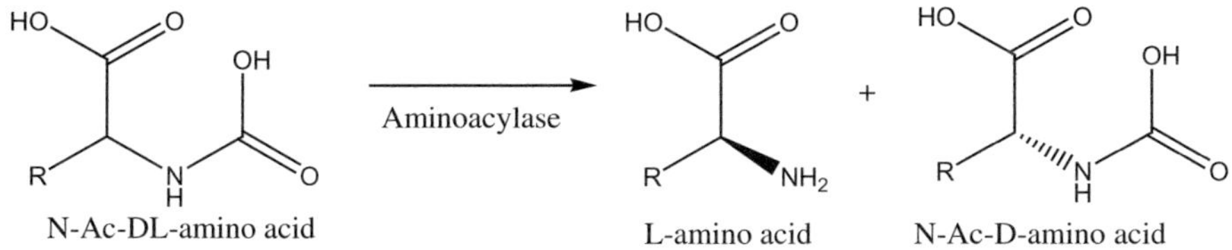

Fig. 10.7 Aminoacylase catalyzed reaction for N-Ac-DL-amino acid to produce L-amino acid

Aspergillus oryzae to produce L-methionine with several hundred tons scale annually (Figure 10.7).

Most substrates are nonnatural in organic and industrial chemistry while enzymes typically show high enantioselectivity towards their natural substrates. Thus, enhancing enzymatic enantioselectivity toward the artificial substrates is necessary for creating superior practical biocatalysts. In the HRP-catalyzed reactions, it can catalyze the oxidation of a variety of chiral phenols with hydrogen peroxide, with low enantioselectivity. To solve the problem, directed evolution has been combined with fluorescence-activated cell sorting (FACS) for improving the HRP's enantioselectivity toward phenols, and the mutant enzyme with a single point mutation showed an 18-fold greater enantioselectivity than the native enzyme (Antipov *et al.* 2009) (Table 10.2).

Case Study: Enantioselective Enzymes by Computational Design (Wijima *et al.* 2015)

Enzyme's enantioselectivity depends on the geometry of the enzyme's structure. As an example, for subtilisin-catalyzed transesterification reaction between vinyl butyrate and S-, R-enantiomers of chiral secondary alcohols was examined using computer-assisted modeling. Higher reactivity of S enantiomer can be explained using steric hinderance as well as the orientation in the active site of the enzyme (Fitzpatrick *et al.* 1992).

A successful strategy to improve enantioselectivity was suggested recently by utilizing molecular dynamics (MD) simulation and epoxide hydrolase-catalyzed conversion of cyclopentene oxide to (R,R)- or (S,S)-cyclopentane-1,2-diol as a model system (Wijma *et al.* 2015). First, the substrate was placed in a desired orientation for enantioselective catalysis in the active site. The orientation is based on the transition state for each enantiomer found by quantum mechanical modeling (Hopmann *et al.* 2005). The active site was then redesigned to stabilize the orientation, via the RosettaDesign method (Hilvert 2013). Eleven positions in the active site were mutated in silico to one of the nine hydrophobic residues (AFGILMPVW) and RosettaDesign found protein sequences appropriate to increase the binding affinity between the substrate and the surrounding active site. These should correspond to the variants with the active site complementary to the substrate in the desired prochiral configuration with low energy. The method generated 236 and 230 pro-SS and pro-RR enzyme structures respectively.

Table 10.2 Enantioselective enzyme processes in Europe (from Schmid *et al.* 2002)

Strategy	Product(s)	Substrate(s)	Catalyst(s)
Kinetic resolution	(*S*)-2-Chloropropionic acid	Racemic 2-chloropropionic acid	Whole cells, (*S*)-specific dehalogenase
	Enantiopure alcohols	Racemic alcohols	Lipases
	4-endo-hyroxy-2-oxabicyclo[3.3.0]-aoct-7-en-3-one	4-hydroxy-2-oxabicyclo[3.3.0]-oct-7-en-3-one butyrate ester	Triacylglycerol acylhydrolase
	Various α-amino acids	Lactams, *N*-protected racemic α-amino acid ester	Lactamases
	Various D-amino acids	Racemic *N*-acylated amino acids	D-Aminoacylase
	(*2R,3S*)-3-(*p*-methoxyphenyl) glycidyl methylester	Racemic trans-3-(*p*-methoxyphenyl) glycidyl methylester	Lipase
	(*S*)-2-phenylpropionic acid	Racemic-2-phenylpropionitrile	Whole cells, Nitrile hydratase, Amidase
Dynamic resolution	(*R*)-mandelic acid	Racemic cyanohydrins	Nitrilase
	Various (*S*)-ester amides	Racemic aralactones	Immobilized triacylglycerol acylhydrolase (triacylglycerol lipase)
	Enantiopure L-amino acids	Racemic *N*-acetyl amino acids	L-Acylases
	Enantiopure D-amino acids	Racemic hydantoins	Hydantoinases, Decarbamylases

The method however, can also generate mutants with undesired substrate poses, thus a screening step is required to reduce such occurrences. Computational screening by high-throughput-multiple independent MD simulation (HTMI-MD) was carried out twice in order to rank the designed variant structures. In this step, the fraction of time that the enzyme-substrate complex spends in a pro-RR or pro-SS conformation over MD timescale was calculated as an *in silico* measure of reactivity. The ratio of the *in silico* reactivity between the enantiomeric product pair would be an indication of *in silico* enantioselectivity. The geometric criteria for the pro-RR or pro-SS conformations are near attack conformations (NACs), which are the conformations that approach the transition state as indicated by quantum mechanical modeling. This includes key interatomic distances smaller than the van der Waals contact distances, the bond angles deviating less than 20° from those of the transition state, and all hydrogen bonds present as in the transition state.

The MD simulations indicated the initial designs could catalyze the reactions for desired enantiomers, but without exclusive selectivity for the enantiomer because there was too much space in the active site which allowed movement of the substrate into both pro-RR and pro-SS orientations. Additional discriminating power between the enantiomers was introduced by steric hindrance from specific bulky mutations, at L103, L114, I116, and F134 to block pro-RR orientation, and at L35, L74, M78, I80, and L103 to block pro-SS orientation. With the introduction of the steric hindrance, many more variants presented the desired enantioselectivity (twofold increase and tenfold increase for pro-SS and pro-RR orientations, respectively.

The method was validated experimentally by testing 10 and 27 pro-RR and pro-SS variants respectively, with the highest ranks from HTMI-MD. 28 out of 33 active variants exhibited the desirable enantioselectivity, and nine of them were highly selective with an *ee* over 75%. The most enantioselective variant had an *ee* of 85.5% for RR and 90.2% for SS. These results show there is a good correlation in enantioselectivity between the computationally screened and the experimentally observed (Fig. 10.8).

Further Discussion

1. How can we explain the differences of enzyme's function for different substrate of the same enzyme?
2. What is the RosettaDesign method and advantages of using RosettaDesign in engineering enzymes?

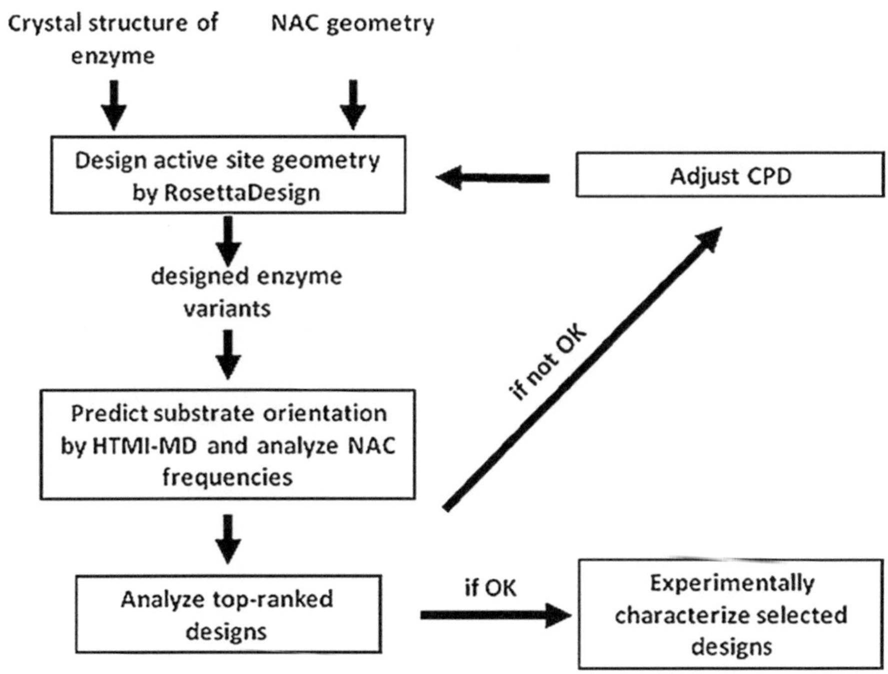

Fig. 10.8 Scheme for designing the active site based on geometry of the substrate for prochiral orientations (Wijima *et al.* 2015) * NAC: near attack conformations, CPD: computational protein design

3. How can we design the pocket geometry of the enzyme's active site for different substrates?
4. Why some enzymes can react both forms and some enzymes act only one form of substrate in enantioselectivity? Discuss this difference in active site geometry and other factors.

References

Andrews FH and McLeish MJ. Using site-saturation mutagenesis to explore mechanism and substrate specificity in thiamin diphosphate-dependent enzymes. *FEBS J*, 2013, 280:6395–6411.

Antipov E, Cho AE and Klibanov AM. How a single-point mutation in horseradish peroxidase markedly enhances enantioselectivity. *J. Am. Chem. Soc.*, 2009, 131:11155–11160.

Arnold FH and Volkov AA. Directed evolution of biocatalysts. *Curr Opin Chem Biol*, 1999, 3:54–59.

Carey FA and Sundberg RJ. Advanced organic chemistry part A: structure and mechanisms. *New York, Plenum Press*, 1984.

Fitzpatrick P, Ringe D and Klibanov A. Computer assisted modeling of subtilisin enatioselectivity in organic solvent. *Biotech.Bioeng*, 1992, 40:735–742.

Gao X, Huang F, Feng J, Chen X, Zhang H, Wang Z, Wu Q and Zhu D. Engineering the meso-diaminopimelate dehydrogenase from *Symbiobacterium thermophilum* by site saturation mutagenesis for d-phenylalanine synthesis. *Appl Environ Microbiol*, 2013, 79:5078–5081.

Gordon SR, Stanley EJ, Wolf S, Toland A, Wu SJ, Hadidi D, Mills JH, Baker D, Pultz IS and Siegel JB. Computational design of an α-gliadin peptidase. *J Am Chem Soc*, 2012, 134:20513–20520.

Hilvert D. Design of protein catalysts. *Annu. Rev. Biochem.*, 2013, 82:447–470.

Hopmann KH, Hallberg BM and Himo F. Catalytic mechanism of limonene epoxide hydrolase, a theoretical study.*J. Am. Chem. Soc.*, 2005, 127:14339–14347.

Keith JM, Larrow JF andJacobsen EN. Practical considerations in kinetic resolution reactions. *Advanced Synthesis & Catalysis*, 2001, 343(1):5–26.

Khare SD, Kipnis Y, Greisen PJ, Takeuchi R, Ashani Y, Goldsmith M, Song Y, Gallaher JL, Silman I, Leader H. Computational redesign of a mononuclear zinc metalloenzyme for organophosphate hydrolysis. *Nat Chem Biol*, 2012, 8:294–300.

Kiss G, Çelebi-Ölçüm N, Moretti R, Baker D and Houk KN. Computational enzyme design. *Angew Chem, Int Ed*, 2013, 52:5700–5725.

Manna SK and Mazumdar S. Tuning the substrate specificity by engineering the active site of cytochrome P450cam: A rational approach. *Dalton Trans*, 2010, 39:3115–3123.

Mouratou B, Kasper P, Gehring H and Christen, P. Conversion of tyrosine phenol-lyase to dicarboxylic amino acid β- lyase, an enzyme not found in nature. *J Biol Chem*, 1999, 274:1320–1325.

Murphy PM, Bolduc JM, Gallaher JL, Stoddard BL and Baker D. Alteration of enzyme specificity by computational loop remodeling and design. *Proc Natl Acad Sci U S A*, 2009, 106:9215–9220.

Otten LG, Hollmann F and Arends IW. Enzyme engineering for enantioselectivity: from trial-and-error to rational design? *Trends in Biotechnology*, 2009, 28(1):46–54.

Rouhi M. Chiral Chemistry: Traditional methods thrive despite numerous hurdles, including tough luck, slow commercialization of catalytic processes. *Chemical & Engineering News*, 2004, 82:47–62.

Schmid A, Hollmann F, Park JB and Bühler B. The use of enzymes in the chemical industry in Europe. *Current Opinion in Biotechnology*, 2002, 13:359–366.

Sinclair R, Reid GA and Chapman SK. Re-design of *Saccharomyces cerevisiae* flavocytochrome b2: Introduction of L-mandelate dehydrogenase activity. *Biochem J*, 1998, 333:117–120.

Sobolev V, Sorokine A, Prilusky J, Abola EE and Edelman M. Automated analysis of interatomic contacts in proteins. *Bioinformatics*, 1999, 15:327–332.

Strauss UT, Felfer U and Faber K. Biocatalytic transformation of racemates into chiral building blocks in 100% chemical yield and 100% enantiomeric excess. *Tetrahedron-Asymmetry*, 1999, 10(1):107–117.

Tyagi S and Pleiss J. Biochemical profiling in silico – Predicting substrate specificities of large enzyme families. *J. Biotechnology*, 2006, 124:108–116.

Tyka MD, Jung K and Baker D. Efficient sampling of protein conformational space using fast loop building and batch minimization on highly parallel computers. *J Comput Chem*, 2012, 33:2483–2491.

Watanabe S, Kodaki T and Makino K. Complete reversal of coenzyme specificity of xylitol dehydrogenase and increase of thermostability by the introduction of structural zinc. *J. Biol. Chem*, 2005, 280:10340–10349.

Wijma HJ and Janssen DB. Computational design gains momentum in enzyme catalysis engineering. *FEBS J*, 2013, 280:2948–2960.

Wijma HJ, Floor RJ, Bjelic S, Marrink SJ and Baker D. Enantioselective enzymes by computational design and *in silico* screening. *Angewandte Chemie*, 2015, 127:3797–3801..

Xie T, Song B, Yue Y, Chao Y and Qian S. Site-saturation mutagenesis of central tyrosine 195 leading to diverse product specificities of an α-cyclodextrin glycosyltransferase from *Paenibacillus sp*. 602–1. *J Biotechnol*, 2014, 170:10-16.

Yeon YJ, Park HY and Yoo YJ. Enzymatic reduction of levulinic acid by engineering the substrate specificity of 3-hydroxybutyrate dehydrogenase. *Bioresource Technology*, 2013, 134:377–380.

Chapter 11
Thermodynamics and Stability

There are many advantages of enzymes such as substrate specificity, mild reaction conditions. However, there are also disadvantages such as high cost and instability of enzymes which give limitation for commercial applications. How can we overcome such disadvantages? By genetic engineering methods, over-production of enzymes becomes possible and cost of enzymes can be finally lowered. Practical application of enzymes as industrial scale requires stable enzymes at high temperature, pH extremes, and even at high concentration of salts, alkalis, and surfactants. For example since in starch gelatinization, the temperature would be around 100 °C and in textile desizing, the temperature range is from 80 to 90 °C, enzymes tolerant at high temperature is therefore especially useful. Enzymes will be subjected to high salt concentration and in the presence of surfactants if they are used in food industry and detergents industry, respectively. Potential employment of enzyme will be feasible in the field of enzymatic analysis for clinical, industrial, and environmental samples. These enzyme biosensors should be stable and easily calibrated. There is a need to stabilize enzyme not just to enhance its potential for industrial application but also for economic reasons. For unstable activity issues, finding or obtaining highly stable enzymes, improvement of operating methods, or formulation of the enzyme solution, etc., can be a solution.

11.1 Enzyme Stability

11.1.1 Understanding of Enzyme Stability

During the life cycle of enzymes including enzyme manufacturing, storage and its utilization, enzymes undergo denaturation. Denaturation of protein is defined as

Y.J. Yoo *et al.*, *Fundamentals of Enzyme Engineering*,
DOI 10.1007/978-94-024-1026-6_11

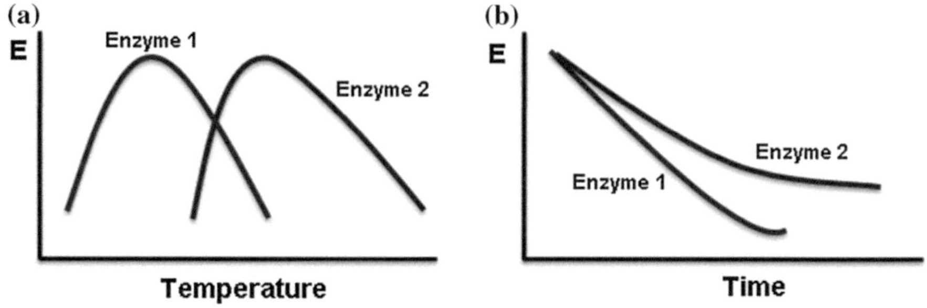

Fig. 11.1 Enzyme activity profile regarded on **a** thermodynamic (conformational) stability and **b** kinetic (operational) stability

the reversible phenomena that key residues of protein are not positioned appropriately close enough for keeping participation in functional and stable structure any more due to disordered state of polypeptide. However, protein subjected to severe chemical alternation over threshold leads to an irreversible inactivation which is determined by complete loss of activity (Fagain 1995).

These unfolding events of polypeptide bring about two definitions of protein stability namely, thermodynamic stability and kinetic stability. The resistance of the folded protein conformation to unfolded one is denoted by thermodynamic stability. On the while, kinetic stability concerns the persistence of enzyme activity as shown in Fig. 11.1(b). In terms of thermodynamic stability and kinetic stability, enzyme number 2 is more stable compared to enzyme 1. These resistance types can be formulate as following;

$$\boldsymbol{N} \overset{K}{\leftrightarrow} \boldsymbol{U} \overset{k}{\rightarrow} \boldsymbol{I}, \tag{11.1}$$

where N, U, and I represents the native, unfolded, and inactivated state of proteins.

(1) **Environmental Factors**

Heat. Heat energy allows a protein structure more flexible and exposes the inner hydrophobic phases to the surrounding solvents, which is reversible to a certain level. However, when the temperature increases over a threshold, hydrogen bonds within α-helices and β-sheets begin to break down. The amide nitrogen and carbonyl oxygen within peptide bonds that used to form the hydrogen bonds in the secondary structures instead form hydrogen bonds to surrounding water molecules, making the process irreversible.

Organic Solvent. The effect of organic solvent on protein stability is dependent on the intrinsic characteristics of the solvent and the ratio of aqueous solvent mixture. In water/water miscible organic solvent, denaturation is fast as the solvent replaces water in hydrogen bonding to the protein. In water/water-immiscible solvent, specific protein structural changes occur at the interface of the two solvents. For example, many lipases require certain structural changes mediated by the

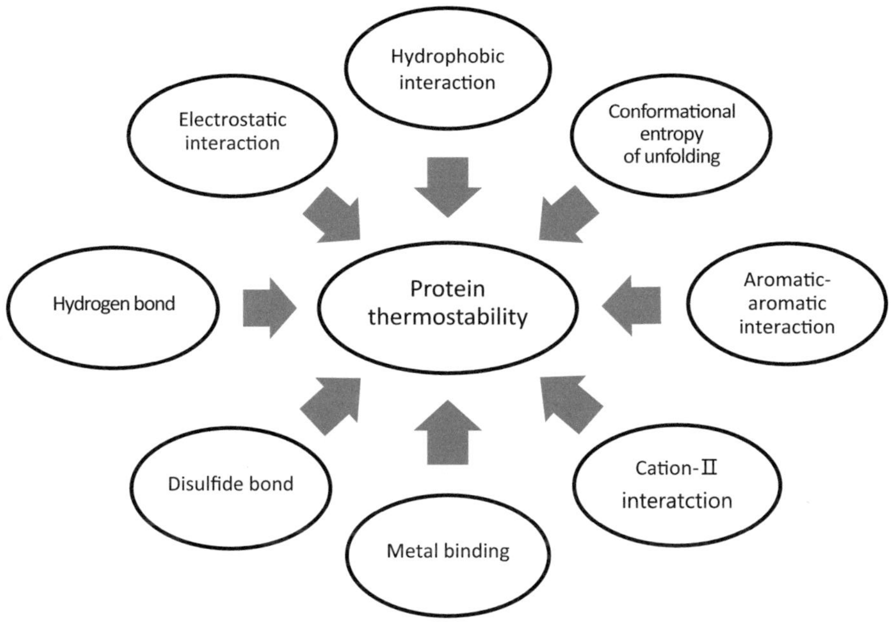

Fig. 11.2 Major factors contributing to stability of a folded protein (Joo and Yoo 2009)

solvent interface for activation. In nearly dry organic solvent system, the degree of polarity of the solvent determines the enzyme activity and stability. As a rule of thumb, the surface sites of enzyme molecules can be replaced with hydrophobic residues to stabilize the enzyme in organic solvent, by forming a water-shell surrounding the enzyme.

pH. Proteins are most stable at pH near the proteins' isoelectric points. Most amino acid residues have a tendency to be charged with positive one at low pH and negatively charged at high pH, in both cases leading to increased electrical repulsion and instability.

(2) **Intrinsic Factors**

Thermodynamic stability of proteins can be affected by several factors as shown in Fig. 11.2 (Kumar *et al.* 2000 and Creighton 1997).

Hydrophobic Interaction. Hydrophobic residues of an enzyme molecule are exposed and come to contact with water molecules when proteins are unfolded, leading to ordering of the water molecules on the surface of proteins. In folded proteins, however, the hydrophobic residues are usually buried in the core of proteins in folded proteins by forming hydrophobic patches to avoid direct touch with surrounding water molecules. Bunching up of hydrophobic residues helps to stabilize a protein by lowering potential energy during protein folding. The hydrophobic interaction expelling water molecules from protein has been recognized as powerful driving force for folding.

Hydrogen Bonding. This is one of the dominant factors that stabilizes α-helices and β-sheets of protein and finely tunes the water–protein interactions. Hydrogen bonds form when significant differences in the electronegativity of atoms, dictated by the Pauling rule, result in the pulling down of electrons and leave the protons partially unshielded (Jeffrey 1997).

Conformational Entropy. A protein harbors a specific conformation through the loss of conformational entropy. For example, glycine residue of one protein shows the largest conformational entropy because it carries only one hydrogen atom in the side chain, which gives more freedom of movement. On the contrary, proline residue has few conformations, which leads to the lowest conformational entropy. Substitution of a glycine residue in target protein to residues with less conformational entropy can stabilize the folding of a protein to a specific conformation (Watanabe *et al.* 1997).

Electrostatic Interactions. Electrostatic interaction found on enzyme is one of common ionic interactions between oppositely charged residues. This interaction is affected by the interatomic distance and the dielectric constant of solvents. The electrostatic interactions get stronger when the distance is decreased, especially forming *salt bridges* or *ion pairs* under 4 Å. The electrostatic interactions are also stronger when there are large differences in dielectric constants of the solvents, such as in a nonpolar medium rather than in water.

Disulfide Bonds. These are made between two oxidized cysteine side chains. Disulfide bonds can be found within single polypeptide chain as well as between multiple polypeptides. It helps to stabilize folded proteins by an entropic effect, where the entropy of the unfolded protein is reduced since disulfide bonds exist covalently even in the unfolded state.

Aromatic–Aromatic Interaction and Cation–π Interaction. This is caused by the intimate interaction between aromatic residues (e.g., Phe, Tyr, and Trp). In addition, cation–π interaction exists where noncovalent bonds between aromatic residues and positively charged residues such as Arg, Lys, and His occur. Perpendicular positioning of these two interactions within 6 Å of the aromatic ring centroids is the most favored geometry found to facilitate the stabilization of protein marginally (Ma and Dougherty 1997).

Investigation on thermophilic enzymes has revealed many characteristics such as more hydrophobic residues, more disulfide bonds and more ionic interactions compared to mesophilic enzymes. Psychrophilic enzymes have shown more flexible characters by having small and easily movable amino acids such as glycine. Sometimes, metals are important in stabilizing enzymes such as calcium in α-amylase (Fig. 11.2).

11.1.2 Measurement of Stability

Proteins under the reversible folding process, with structural changes, can be monitored through the changes in the protein's optical property, viscosity, light

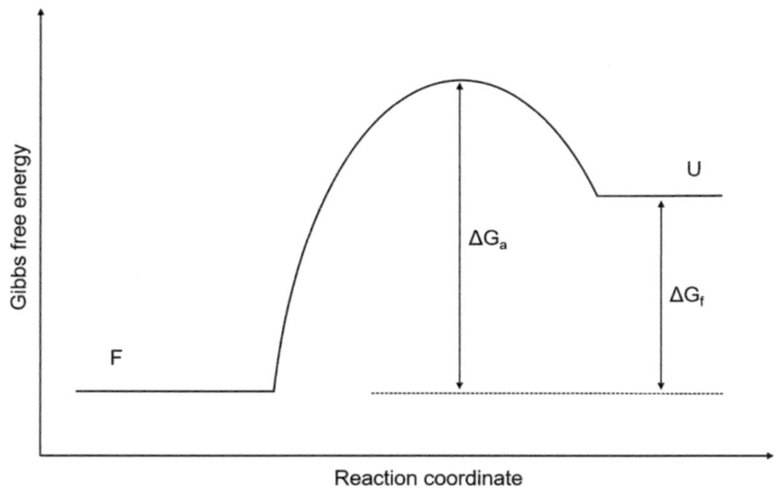

Fig. 11.3 Schematic demonstration of energetics regarding on protein folding procedures. **a** Proteins are stabilized by lowering the potential energy of the folded state (F) or increasing the potential energy of the unfolded (U) or through combined approaches. **b** Kinetic stability of proteins can be improved by increasing ΔG_a. F, folded state; U, unfolded state; ΔG_f, Gibbs free energy difference between folded and unfolded states; ΔG_a, activation free energy of unfolding

scattering property, and turbidity. This thermodynamic (or conformational) stability of proteins can be formulated as the difference of Gibbs free energy between the folded state and the unfolded state using the following equation,

$$\Delta G_f = G_f - G_u, \tag{11.2}$$

where ΔG_f, G_f, G_u refers to the Gibbs free energy difference between folded and unfolded, Gibbs free energy of the folded state and Gibbs free energy of the unfolded state, respectively. ΔG_f will be negative if folding were thermodynamically favored (Fig. 11.3).

Between native and engineered proteins, the difference in thermodynamic stability [$\Delta\Delta G_f = \Delta G_f$(native)—$\Delta G_f$(mutant)] can be determined. Thermodynamic stability of the target protein can be measured by various methods including circular dichroism (CD), differential scanning calorimeter (DSC), and calorimetry. The reversibility of protein unfolding from Eq. 11.1 implies that the folded protein is in equilibrium with unfolded molecule. For small proteins, the equilibrium between the folded (N) and unfolded (U) form is described as follows:

$$N \overset{K}{\leftrightarrow} U \quad \text{and} \quad K = \frac{U}{N} = \frac{F_N}{1 - F_N} \tag{11.3}$$

$$\text{so that} \quad \Delta G_f = -RT \ln K \tag{11.4}$$

,where F_N and ΔG_f are the fraction of unfolded protein and Gibbs free energy change of unfolding, respectively. The more negative ΔG_f will be the less likely

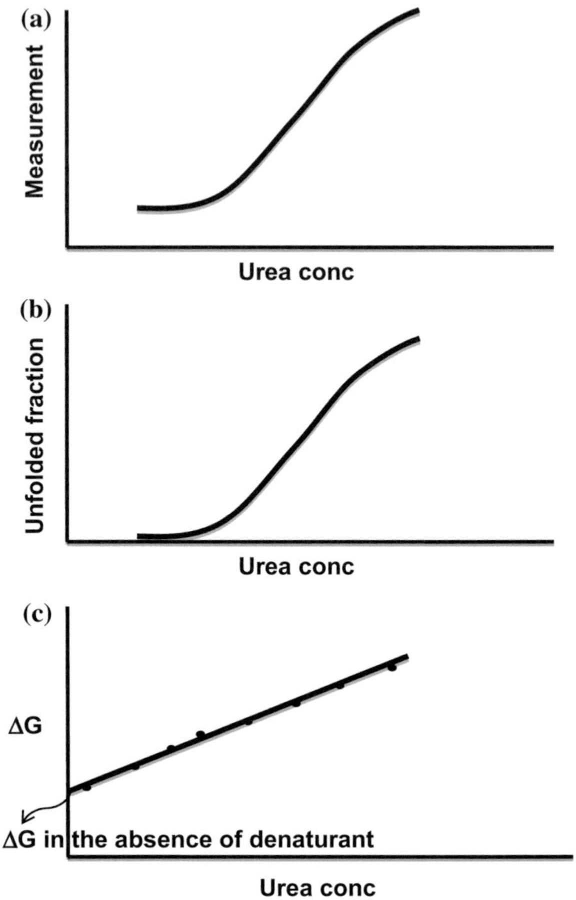

Fig. 11.4 Method to get free energy of unfolding **(a)** measurement, **(b)** calculation of unfolded fraction, **(c)** getting ΔG in the absence of denaturant

is the presence of unfolded molecules. Once protein becomes unfolded, properties such as optical property, viscosity, and molecular size changes. Once aggregation occurs during unfolding, molecular weight also changes, Experimental data such as fluorescence versus urea concentration can be transformed to the plot unfolded fraction versus urea concentration.

The equilibrium constant can be calculated from $K = \frac{F_N}{1-F_N}$ and then the free energy in the absence of denaturant can be estimated. Using this analysis, the obtained experimental data in Fig. 11.4a can be transformed into an estimated value curve of ΔG_f when the urea (denaturation reagent) concentration as independent variable.

The free energy change in the absence of denaturant is not given directly and must be calculated through extrapolation from Fig. 11.4c. For many small proteins, the experimentally determined plot of ΔG_f versus denaturant concentration is approximately linear. Consequently, the simplest way to estimate ΔG in the

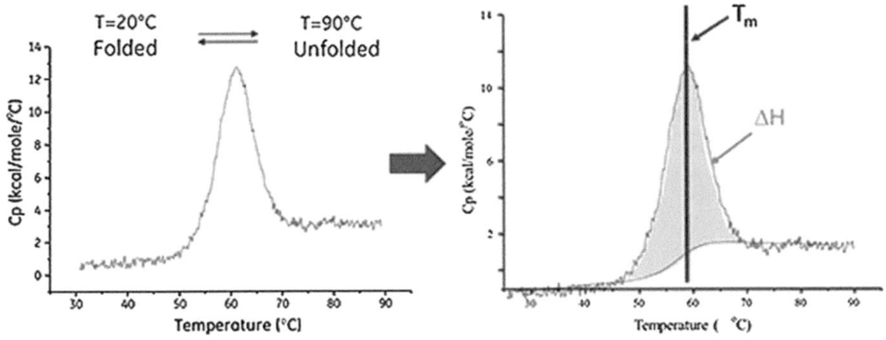

Fig. 11.5 Conceptual explanation of DSC data for T_m and enthalpy

absence of denaturant is to assume that this linear dependence continues to zero concentrations.

Thermodynamic stability. One of the analyzing the thermodynamic stability for a protein dissolved in dilute solution requires the determination of changes in the molar heat capacity of concerned protein under constant pressure condition (ΔC_p). Heat capacity of a protein molecule mirrors its capability to absorb heat and expresses increased tendency in temperature. Heat energy is taken in by the protein and lead to unfold state over a temperature range and bring about a characteristic endothermic peak for each protein (Fig. 11.5). During the formation of unfolded state, water molecules around the protein rearrange since more part of hydrophobic residues of protein are revealed. Once unfolding is over, heat absorption declines and a baseline is newly organized. Integration of the heat capacity of the sample versus temperature yields the enthalpy (ΔH) of the unfolding process, which is caused by breaking of hydrogen bonds and disruption of hydrophobic interactions found in proteins (Ladburg *et al.* 1995).

$$\Delta \mathrm{H} = \int_{T1}^{T2} C_p \mathrm{d}T \tag{11.5}$$

The transition midpoint (T_m) can be defined as the temperature at which half of the proteins are assumed folded and the other unfolded state. The entropy (ΔS) term is estimated from the integrated area under the line of C_p/T versus T.

$$\Delta S = \int_{T1}^{T2} (C_p/T) \mathrm{d}T \tag{11.6}$$

Differential Scanning Fluorimetry (DSF).

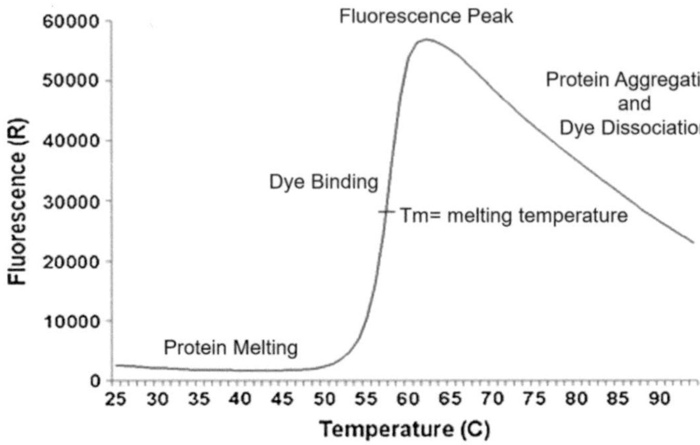

Fig. 11.6 Typical DSF analysis result for determination of T_m of target protein

Recently new method to estimate Tm based on fluorimetry has been suggested since relatively simple and cheap flurospectrometry is more available compared to DSC. Fluorescent dye such as SYPRO Orange makes nonspecific bonding with hydrophobic exposed patch of protein, on the while water molecules strongly extinguishes dye-protein fluorescence. Protein unfolding facilitates the exposed surfaces to bind more fluorescent dye molecules, which result in an enhanced signal in fluorescence by excluding water. The midpoint value on the curve between fluorescence and temperature is acquired by gradually stepping up the temperature up to peak point to unfold the protein (Fig. 11.6).

11.2 Stabilization of Enzyme

11.2.1 Enzyme Stabilization

As shown in Fig. 11.7, three kinds of methods can represent and be utilized for protein stabilization. The first method is to screen better enzymes among the various kinds of qualified enzymes or randomly DNA molecules from environmental metagenomic libraries. The second method is the typical approach to modify protein's environment such as freeze-drying, adding stabilizing agents and an immobilization on carriers. The third method is protein engineering in which proteins are genetically modified.

For traditional approach, polyethylene glycol derivatives, functionalized β-cyclodextrin derivatives have commonly been used. Chemical modification can increase stability of enzyme in organic solvents or even thermostability. Added ingredients such as polyol and fumed silica were reported to give a stabilizing impact to enzymes. Protein immobilization can give higher stability towards organic solvents and temperature.

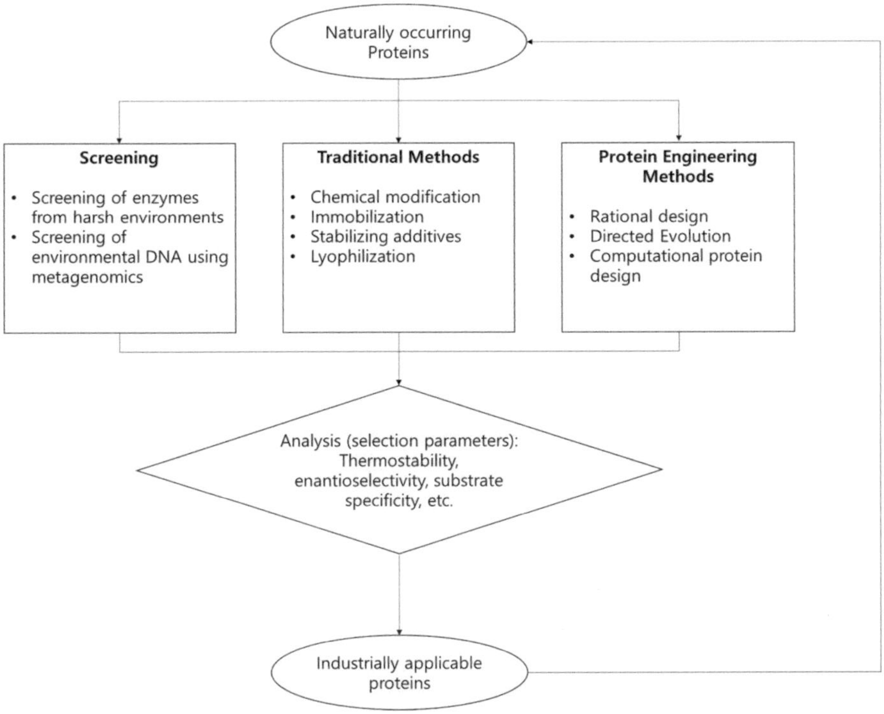

Fig. 11.7 General approaches to improve protein stability (Joo and Yoo 2009)

Screening. Extremophiles whose habitats are characterized by high temperature, pressure, pH, or salt concentration such as volcanic vents, hot spring, or deep sea are useful for isolating novel enzymes that are stable in such environment. From thermophile, psychrophile, acidophile, alkalophile, halophile, and barophile, various enzymes that are stable for industrial applications were identified. For example, the enzymes used in saccharification of starch are from thermophilic microorganisms.

Additives. Various additives such as sugar, amino acid, detergent, fatty acid, metal, polyol, antioxidants are used to enhance the stability of enzymes. For example, acidic amino acids are used to prevent loss of lysozyme by inhibiting the adsorption of the enzyme to glass instruments. Polyols such as sorbitol, glycerol, and mannitol are commonly used in the freeze-drying of proteins. Salts and antioxidants are important components of protein drug formulations.

Immobilization. Enzymes can be immobilized on organic and inorganic carriers to increase their half-life. It is important to increase the amount of protein immobilized per g carrier, and doing so without losing significant enzyme activity. However, this approach is still at trial-and-error level, and researches to predict which carrier material is optimal for a specific enzyme and how to retain activity while increasing stability are ongoing.

Chemical Modification. Certain amino acid residues on the enzyme surface are capable of chemical modifications such as glycosylation, methylation, and PEgylation. Enzymes can also be subjected to site-directed mutagenesis to introduce such residues if there were none on the surface. PEgylation, where polyethylene glycol (PEG) with both hydrophobic/hydrophilic properties and variable molecular weight is attached covalently to the residues, is often used to modify the polarity of the enzyme surface. Researches are being carried out on various PEG substitutes.

Solvent Engineering. Specific solvents at their various concentrations lead to variable water activity, and enzyme stability and activity are thus changed. Currently, trial-and-error method is mostly used to try different solvents and see if the stability or activity is increased.

Protein Engineering. By mutating the amino acid residues in proteins, intramolecular interactions can be enhanced to increase stability. There are mainly three methods to induce mutation in protein structures.

First, rational design relies on protein structural information. Specific motifs such as proline, disulfide bond, hydrophobic cluster, hydrogen bonding are introduced at appropriate locations in the protein structure via site-directed mutagenesis.

Second, directed evolution creates a random mutant library and uses high-throughput selection methods to choose stability-enhanced mutants.

Third, computational design is similar to the rational design in terms of relying on protein structural information and site-directed mutagenesis, but with more help from computational tools such as docking, molecular dynamics simulation, and quantum mechanics-molecular mechanics simulation. Aside from rational design and directed evolution, recently de novo design through computational methods is getting attention for protein design. For de novo design, optimal polypeptide sequences must be searched to be fit with backbone of target protein.

In principle, increase of ionic interaction, increase of hydrophobicity in the core, shortening surface loops, increase proline content, increase disulfide bonds where possible are examples to increase the thermal stability.

Nowadays, combinations of the three are used rather than a single method to enhance the stability of enzymes. And domain-based approach to find efficient mutation strategy is desirable instead of aiming whole globular protein or neglecting 3–5 Å boundary from the active site.

11.2.2 Applications to Enzyme Pharmaceuticals

It is very important to maintain the activity of protein drug during storage. For this purpose, many researches have been performed. Protein drug including enzyme drug is to be analyzed for its safety, for example, molecular sizes, activity.

Formulation development is thus important and should consider also delivery method such as injection, aerosol, etc. (Shire 1996). For the formulation, mannitol, glycine, or other alcohols with buffering agents are usually added to protect the protein from denaturation and aggregation during storage. The proteins tend to aggregate when stored, which affect the activity and immunological effect. Through molecular simulation study, protein region for potential aggregation can be predicted. Based upon the prediction, engineering of the protein for enhanced stability becomes possible (Chennamsetty *et al.* 2009).

Another approach is to modify the enzyme. Examples are modification using PEG (Polyethylene glycol) and making conjugates with other protein. Enzyme-HSA (Human Serum Albumin) can be an example to increase the in vivo stability of the therapeutic enzyme.

Case Study: Structure-Based Pattern Analysis for Protein Stability (Pack and Yoo 2003)

One critically important factor "weak thermostability" may impede the application of enzymes as useful biocatalysts in various industrial sections. This prompted biophysical and biotechnological efforts to understand the main reason and mechanism for weak thermostability of enzymes. Enzymes showing thermophilic properties retain the common fold shared in the peculiar protein family. Comparative studies compared the loop stability, compactness of thermophilic enzymes with their mesophilic proteins. However, no clear different characteristics in structures were found due to intrinsic limitation such as (1) similar common folds found both on thermophilic and mesophilic microbial and (2) no available analysis method capable of discriminating the details.

Pack and Yoo (2003) reported the comparison about the distribution and residue structure distinctly found between 20 families of thermophilic and mesophilic counterparts. Table 11.1 shows the data set used in the investigation (Table 11.2).

Structural differences of amino acids can give effect on protein stability. Aliphatic amino acids including ALA may play a major role in the hydrophobic interaction, which is the major interaction to maintain conformational stability or core shield. Positive or negative charge containing amino acids would form electrostatic bonds, which is crucial interaction for conformational stability at the surface part. On the while, aromatic amino acids can form cation–pi interaction, which can contribute structural stability of protein. The information obtained can be used for rational design of enzymes.

Table 11.1 20 different proteins pairwise data set

Protein name	Thermophilic proteins	Mesophilic proteins
	PDB code/organism/temperature (°C)	PDB code/organism/temperature (°C)
Adenylate kinase	1zin/*Bacillus stearothermophilus*/40-65	1aky/*Saccharomyces cerevisiae*/25-30
Che Y	1tmy/*Thermotoga maritima*/80-85	3chy/*Escherichia coli*/37
Citrate synthase	1aj8/*Pyrococcus furiosus*/100	1csh/Chicken heart/37
EF-TS and EF-TU-TS	1tfe/*Thermus thermophilus*/70-75	1efu_b/*Escherichia coli*/37

(continued)

Table 11.1 continued

Protein name	Thermophilic proteins	Mesophilic proteins
	PDB code/organism/temperature (°C)	PDB code/organism/temperature (°C)
Endo-1,4-b xylanase	1yna/*Thermomyces lanuginosus*/50	1xnb/*Bacillus circulans*/30-40
Glutamate dehydrogenase	1gtm/*Pyrococcus furiosus*/75-100	1hrd/*Clostridium symbiosum*/30-37
Glyceraldehyde-3-phosphate dehydrogenase	1hdg/*Thermotoga maritima*/80-85	1gad/*Escherichia coli*/37
Inorganic pyrophosphatase	2prd/*Thermus thermophilus*/70-75	1ino/*Escherichia coli*/37
Lactate dehydrogenase	1ldn/*Bacillus stearothermophilus*/40-65	1ldg/*Plasmodium falciparum*/37
Malate dehydrogenase	1bdm/*Thermus flavus*/70-75	4mdh/*Porcine*/37
Manganese superoxide dismutase	3mds/*Thermus thermophilus*/70-75	1qmn/*Homo sapiens*/37
Methionine aminopeptidase	1xgs/*Pyrococcus furiosus*/100	1mat/*Escherichia coli*/37
Phosphofructokinase	3pfk/*Bacillus stearothermophilus*/40-65	2pfk/*Escherichia coli*/37
3-Phospho glycerate kinase	1php/*Bacillus stearothermophilus*/40-65	1qpg/*Saccharomyces cerevisiae*/25-30
Reductase	1ebd/*Bacillus stearothermophilus*/40-65	1lpf/*Pseudomonas fluorescens*/25-30
Ribonuclease H	1ril/*Thermus thermophilus*/70-75	2rn2/*Escherichia coli*/37
Rubredoxin	1caa/*Pyrococcus furiosus*/100	8rxn/*Desulfovibrio vulgaris*/34-37
Subtilisin	1thm/*Thermoactinomyces vulgaris*/55-65	1st3/*Bacillus lentus*/30
Thermolysin	1lnf/*Bacillus thermoproteolyticus*/53	1npc/*Bacillus cereus*/30
Triose phosphate isomerase	1btm/*Bacillus stearothermophilus*/40-65	1ypi/*Saccharomyces cerevisia*/25-30

Note PDB Code/organism/temperature denotes the Protein Data Bank code, source organisms of the protein and the optimum growth temperature of the source organism

Table 11.2 Structural properties of the thermostable enzymes compared to mesophilic enzymes

Characteristics	Location
Higher frequency of salt-bridge	Exposed location
Lower frequency of flexible residue	Fully-exposed location
Higher frequency of flexible residue	Boundary location
Higher frequency of hydrogen bonds	Well-buried location

Further Discussion

1. What is the relationship between thermodynamic stability and kinetic stability? Does high thermodynamic stability mean high kinetic stability? Can we increase thermodynamic stability and kinetic stability at the same time?
2. What are the basic principles to increase thermodynamic stability and kinetic stability?
3. Gibbs free energy can be expressed using enthalpy and entropy. Discuss stability issues using enthalpy and entropy.
4. It happens frequently that increase of thermal stability resulted in the decrease of the activity. Can we increase the activity and stability of enzymes at the same time?
5. Find the story of isolation of thermophilic DNA polymerase enzymes.

References

Chennamsettty NC, Voynov V, Kayser V, Helk B and Trout BL. Design of therapeutic proteins with enhanced stability. *PNAS*, 2009, 106:11937–11942.

Creighton TE. *Proteins: Structures and Molecular Properties* (second ed.), Freeman & Company, 1997.

Fagain O. Understanding and increasing protein stability. *Biochimica et Biophysica Acta*, 1995, 1252:1–14.

Jeffrey GA. *An introduction to hydrogen bonding*. Oxford University Press; 1997.

Joo JC and Yoo YJ. Thermostable proteins. *Encyclopedia of Industrial Biotechnology: Bioprocess, Bioseparation, and Cell Technology* (edited by Michael C. Flickinger), John Wiley & Sons, 2009.

Kumar S, Tsai CJ and Nussinov R. Factors enhancing protein thermostability. *Protein Engineering*, 2000, 13:179–191.

Ladbury JE, Wynn R, Thomson JA and Sturtevant JM. Substitution of charged residues into the hydrophobic core of *Escherichia coli* thioredoxin result in a change in heat capacity of the native protein. *Biochemistry*, 1995, 34:2148–2152.

Ma JC and Dougherty DA. The cation-pi interaction. *Chemical Reviews*, 1997, 97:1303–1324.

Park SP and Yoo YJ. Protein thermostability: structure-based difference of residual properties between thermophilic and mesophilic proteins. *Journal of Molecular Catalysis B: Enzymatic*, 2003, 26:257–264.

Shire SJ. Stability characterization and formulation development of recombinant human deoxyribonuclease I. in *Formulation, Characterization and Stability of Protein Drugs* (Rodney Pearlman and Y. John Wang eds.), Plenum Press, 1996.

Watanabe K, Hata Y, Kizaki H, Katsube Y and Suzuki Y. The refined crystal structure of *Bacillus cereus* oligo-1, 6-glucosidase at 2.0 Å resolution: structural characterization of proline-substitution sites for protein thermostabilization. *Journal of Microbiology and Biotechnology*, 1997, 269:142–153.

Part IV
Applications of Enzymes

Chapter 12
Enzymes for Chemicals and Polymers

Isolated enzymes have been used as highly specific catalyst in organic chemicals synthesis. However, industrial significance of enzyme reactions was especially emphasized in the 1970s with the production of high fructose corn syrup (HFCS). Recombinant DNA technology also enables the efficient production of enzymes, making them cheaply available for use. The introduction of directed evolution technologies for enzyme improvement and the continuous development of new concepts and technologies in biocatalysis expands the uses of enzymes for chemical and pharmaceutical industries (Drauz and Waldmann 1995). The outstanding properties in the aspect of functional selectivity, regioselectivity, and enantioselectivity made the enzymes suitable for fine chemicals and pharmaceutical applications. Precursors and active ingredients of pharmaceuticals, foods, flavors, etc., are widely synthesized using enzyme technology. Recently, cytochrome P-450 monooxygenase and tyrosinase and many other enzymes have been extensively used and studied for organic synthesis. Nowadays, with the development of metabolic engineering and systems biology, enzymes are becoming more and more important for bio-based chemicals and polymers synthesis. In this chapter the industrial applications of biocatalysis and some promising developments in chemicals and polymers production are emphasized and introduced as well as semisynthetic penicillins and cephalosporins which are also very important and classical issues.

12.1 Chemicals

12.1.1 Semisynthetic Penicillins and Cephalosporins

Penicillin is the first found antibiotic from nature and nowadays it is one of the most bulk produced antibiotics approximately 30,000 tons/year. Penicillin G and penicillin V, which are classified as first generation antibiotics possess still major

Y.J. Yoo *et al.*, *Fundamentals of Enzyme Engineering*,
DOI 10.1007/978-94-024-1026-6_12

commercial and clinical significance. However, even though these compounds are still useful antibiotics, most of these penicillins are nowadays produced for the production of β-lactam intermediates, 6-aminopenicillanic acid (6-APA) and 7-aminodesacetoxycephalosporanic acid (7-ADCA). To further augment the potency of penicillins and broaden their antimicrobial range, the first generation penicillins are also used as precursors for semisynthetic penicillins.

Semisynthetic penicillins (SSPs) are engineered to improve side effect profile to have low toxicity and superior pharmacokinetics. The traditional chemical production of SSPs carries out under exceptionally low temperatures to prevent breaking up of vulnerable β-lactam ring. Blocking reagent such as chlorosilanes are needed to shield the penicillin C(3)-carboxyl. Dilation of 6-APA leads to the target SSP. However, SSPs are routinely synthesized in a two-step mode; a schemes for the conversion of semisynthetic β-lactam is shown in Fig. 12.1. First, penicillins

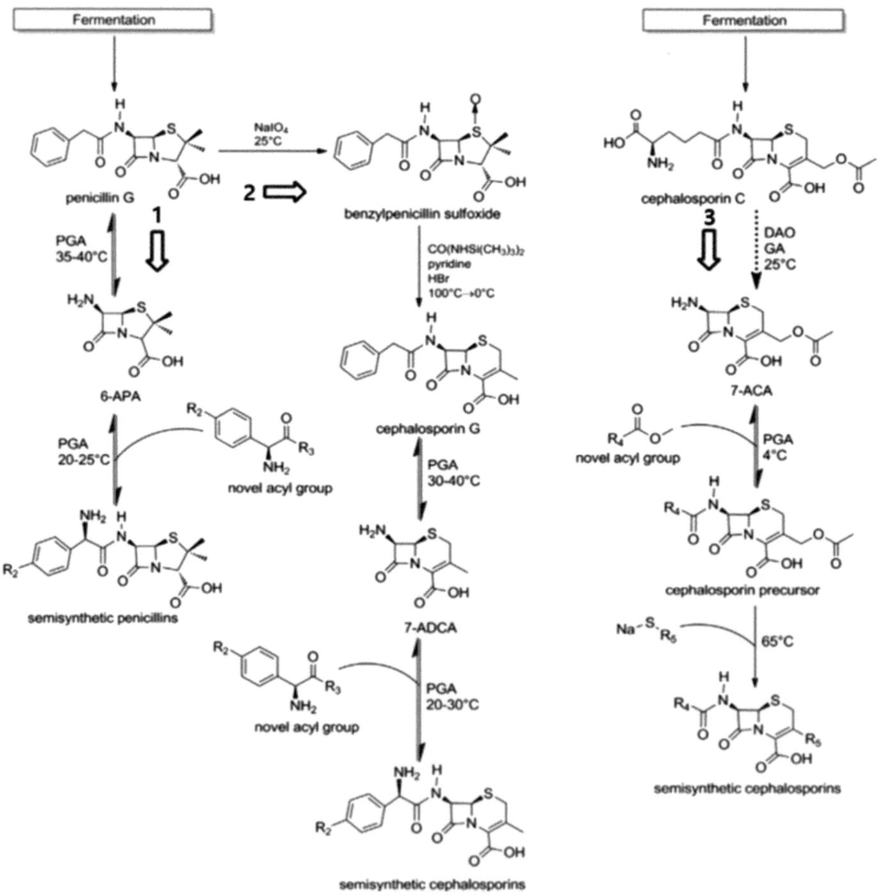

Fig. 12.1 Schematic diagrams for synthetic pathways including enzymatic procedures for the production of semisynthetic penicillins and cephalosporins

are transformed into 6-APA through enzymatic or chemical route. Next, it is further converted into SSPs through enzymatic condensation with the various amide or ester. Biotransformation of 6-APA generally does not need harsh reactants or operating environments (Parmar *et al.* 2000; Wu *et al.* 2010).

6-APA and phenylacetic acid are produced from penicillin G by the enzyme reaction. Since phenylacetic acid is formed, pH of the reaction medium becomes low which affects the enzyme reaction. Control of pH is therefore very important as well as temperature control. For this purpose, recycled reactor system is preferred because of easy control of pH and temperature instead of simple packed bed immobilized enzyme reactor system Fig. 12.2.

Cephalosporins are antibiotics, β-lactam antibiotics. The cephalosporin molecule has the cefem group (linked β-lactam and dihydrothiazine ring) and the penam group (linked β-lactam and thiazolidine ring). The cefem and penam groups determine the antimicrobial activity of these two antibiotics; cleavage of the rings makes therefore loss of antimicrobial properties (Chandel *et al.* 2008).

From Fig. 12.1, cephalosporin G can be chemically synthesized upon oxidative 5-membered thiazolidine ring expansion of penicillin G and consequent elimination of PAA (PATH 2). Penicillin G acylase (PGA) can biotransform cephalosporin G to produce 7-ADCA. The production of 7-ADCA takes place under environmentally benign conditions. Instead semisynthetic cephalosporins can be produced from cephalosporin C as starting chemical through another intermediate of 7-aminocephalosporanic acid (7-ACA) in Path 3 in Fig. 12.1. D-amino acid oxidase and glutaryl acylase are used to enzymatically hydrolyze cephalosporin C to produced 7-ACA. Recently, one-step reaction for 7-ACA synthesis is known to be developed using enzyme technology.

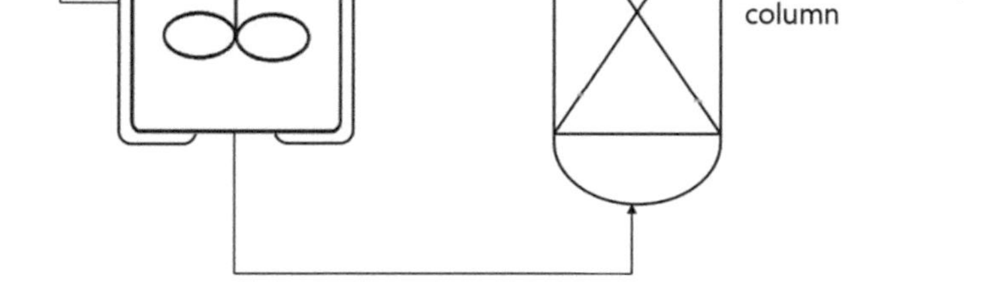

Fig. 12.2 Recycled enzyme reactor for 6-APA from penicillin G **pHC* pH control, *TC* temperature control

12.1.2 Bulk Chemicals

The status of enzymes in the production of bulk chemicals seems very limited but there is significant success story despite the belief that enzymes can be only applied for the synthesis fine chemicals. The belief has background that enzymes are very expensive and limited life span due to low stability compared with conventional chemical catalysts. If enzymes show sufficiently excellent stability and volumetric productivity, of course enzymes can occupy number one priority in the choice of catalysts for bulk chemical synthesis. Recently, protein engineering and novel selection tools based on metagenomic library and in silico genome databases result in the development of exceptionally stable and productive enzymes for bulk chemical conversion. Not only the in vitro usage of enzymes but also in vivo application especially in microbial cell will be viable option in the production of bulk chemicals since key enzymes are being synthesized in the cell by consuming low-cost carbon sources (Table 12.1).

Acrylamide from Acrylonitrile

One of the most successful stories in enzyme technology came from hydration reaction of nitrile group. Polyacrylamide called as PAM in short is a polymer (–$CH_2CHCONH_2$–) from acrylamide as monomer units. It can be polymerized either as a linear or branched structure. PAM shows characteristics such as excellent water absorbing and forms soft gel with water hydration. These gel formation properties render it to be used in gel electrophoresis and soft contact lenses. In addition, recently aesthetic surgery began to employ this PAM as subdermal filler. However, the largest share of PAM can be found to flocculate suspended solid particles in water. This flocculation property is commonly applied in water treatment and papermaking process. Even though PAM is being provided in a powder or liquid formulation, emulsion type PAM can be also found in supplied products. Highly viscous PAM solutions can be manufactured with dosage of little PAM on the spot and injected to oil well to enhance the oil recovery.

The degree of polymerization is known heavily dependent on the purity of monomer since metals and anions can retard the polymerization severely.

Table 12.1 Examples of bulk chemicals production through enzyme catalysis

Enzymes	Reaction	Products	Scale
Nitrile hydratase	Hydration of nitrile group	Acrylamide	30,000 ton/year
		Nicotin amide	3400 ton/year
Aminoacylase	Hydrolysis	Enantio-pure amino acids	50,000 ton/year
Glucose isomerase	Isomerization	HFCS (high fructose corn syrup)	Over several million tons/year
Hydroxylase	Monooxygenation	Herbicide intermediate	Over hundred tons/year
Protease	Peptide bond formation	Aspartame	12,000 ton/year

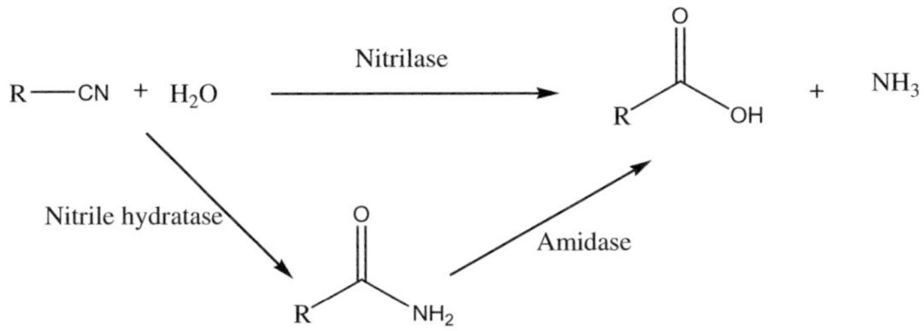

Fig. 12.3 Enzymes involved in the hydrolysis of nitrile compounds

Table 12.2 Various nitrile compounds with reactive concentration by nitrile hydratases (Yamada and Nagawawa 1990)

Amides	Concentration (g/L)	Amides	Concentration (g/L)
Nicotin amide	1465	Difluorobenzylamide	393
Γ-Picolylcarboxamide	1099	2-thienylcarboxamide	254
α-Picolylcarboxamide	977	Indole-3-carboxamide	697
Pyrazine carboxamide	985	Benzylamide	848

Chemical process to produce monomer acrylamide from acrylonitrile has employed Cu based catalyst, which post removal of heavy metal Cu is essential unit process to guarantee the quality of monomer. Hideaki Yamada and his research team in Kyoto University of Japan screened and developed a process based on nitrile hydratase of *Rhodococcus rhodochrous* J1. Nitrile hydratases are harboring Fe^{3+} or Co^{3+} at active center and catalyze the hydrolysis reaction leading to amide synthesis as shown in Fig. 12.3.

On the while nitrilases catalyze the addition of two water molecules to produce carboxylic acid. Acrylonitrile as well as broad range of aromatic nitrile compounds was found to be hydrolyzed by nitrile hydratase. Interestingly exceptionally high concentration of substrate was converted to produce equivalent product by nitrile hydratase (Table 12.2).

In case of liquid nitriles, neat solvent state was achieved during reaction. The extremely high concentration of substrate over 1000 g/L resulted in the very high volumetric productivity, which enabled the industrial application due to low capital cost.

Industrial scale-up was achieved by initially Nitto Chemical with scale of 30,000 ton per year by feeding 25–40% acrylonitrile at cool temperature (around 10 °C) with quantitative conversion as well as over 99.9% of selectivity. Nowadays this enzymatic process became standard process for the production of acrylamide since economic feasibility outcompete conventional chemical process.

Similar process to convert 3-cyanopyridine (nicotine nitrile) for the production of nicotinamide was developed based on same nitrile hydratase as shown in Fig. 12.4.

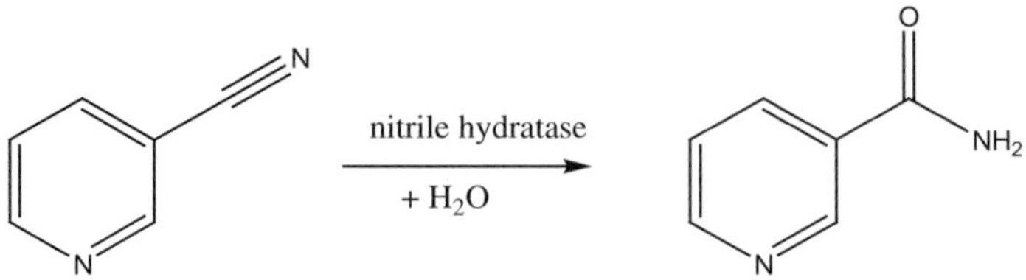

Fig. 12.4 Schematic diagram of hydration reaction of 3-cyanopyridine to nicotinamide (Thomas *et al.* 2002)

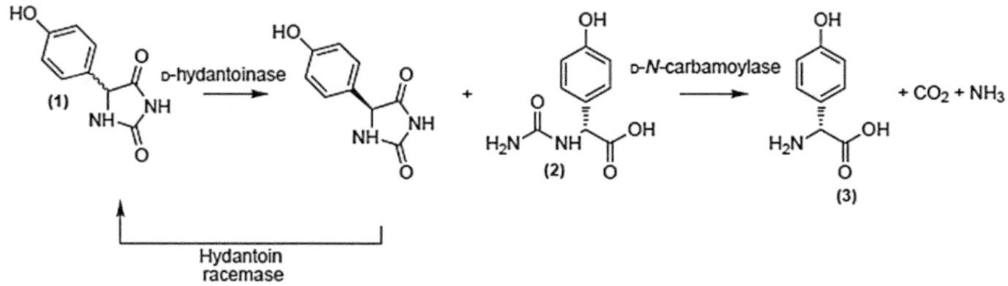

Fig. 12.5 Production of D-*p*-hydroxyphenyl glycine by kinetic resolution by employing both a D-hydantoinase and a D-*N*-carbomoylase (Schulze and Wubbolts 1999)

Nicotinamide finds application as vitamin B_3 since vitamin B_3 has huge demand in live stock feed as well as nutrient for humans and was commercialized by Lonza (Swiss) by using licensed nitrile hydratase from Kyoto University. Lonza in China successfully produce nicotinamide with capacity of 3400 ton per year. The process is known composed of a number of stirred tank type reactors fed with 3-cyanopyridine as precursor. Over 99.3% of yield was reported without producing by-product nicotinic acid, which may cause diarrhea in farm animals. Very selective conversion of this enzymatic process is the key factor to compete well with the chemical process.

Amino Acid Production

D-*p*-hydroxyphenyl glycine has been used for the production of semisynthetic cephalosporins and semisynthetic penicillines. The successful industrial implementation of kinetic resolution was carried out in Ajinomoto, Bayer and DSM. As shown in Fig. 12.5, these enzymes catalyze the hydrolysis of D-*p*-hydroxyphenyl glycine hydantoin, to D-*p*-hydroxyphenyl glycine, via *N*-carbanoyl-D-*p*-hydroxyphenyl glycine on multi-thousand tons scale. The remaining L-*p*-hydroxyphenyl glycine hydantoin spontaneously racemizes to D,L-*p*-hydroxyphenyl glycine hydantoin.

Another example of amino acid is aspartic acid for aspartame. Aspartame showing 200 times sweetness compared with sucrose is being commonly adapted

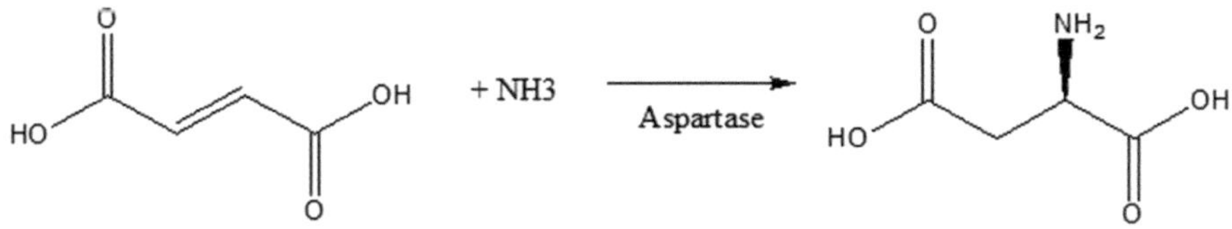

Fig. 12.6 L-aspartate from fumaric acid through enantiospecific addition of ammonia by aspartase

in various food products. The precursor of aspartame was enzymatically synthesized using the enzyme thermolysin. The precursor formation uses the backward hydrolysis reaction of peptide bond. Successful commercialization of diet sweetener Aspartame triggered the demand for L-aspartic acid. Although L-aspartic acid can be produced through aminoacylase process, selective addition reaction using aspartase proved superior since almost quantitative conversion is achieved compared with atmost 50% conversion of acylase reaction. Aspartase from recombinant *Escherichia coli* immobilized on carrageenan gel hardened by glutaraldehyde enabled the production of 1 M L-aspartate on a scale of 4000 ton per year Fig. 12.6.

12.1.3 Enzymes for Metabolic Engineering

In metabolic engineering for synthetic pathway, what enzymes are to be inserted and deleted is a key issue. Liao group (Hanai *et al.* 2007) at UCLA,USA demonstrated the production of isopropanol from *E. coli.* Even though *Clostridium* was known to produce isopropanol, the concentration from this microorganism was very low. They searched the genes from data bases required to synthesize isopropanol, and then the genes were cloned into *E. coli*, since *E. coli* is easier to do metabolic engineering than *Clostridium*. For industrial applications, debottlenecking of the rate-limiting step enzyme, increase the solvent resistance, reduce the product inhibition if exist are required as well as using cheap carbon source to make the product concentration high enough with high yield for commercialization.

Another example of using enzymes in metabolic engineering is the production of 1,4-butandiol (Liu and Lu 2015). 1,4-Butandiol, one of the important industrial chemicals can be produced from D-xylose as a starting carbon source using engineered *E. coli*.

These examples show the importance of enzyme in metabolic engineering. After selecting the source of enzyme from protein data bases, it is also important to improve the enzyme for specific purpose such as substrate specificity, activity. Understanding of enzyme at molecular level is thus required for successful metabolic engineering.

12.2 Polymers

Biocatalysis is being performed until recently under mild conditions like an aqueous solution and at room temperature/atmospheric pressure similar like in vivo condition. However, lift off in the 1980s the concept of biocatalytic reactions under non-natural environment such as organic solvent has been actively started.

The research applied to the field of polymer synthesis utilizing enzymes as active polymerization catalyst has been conducted systemically in many research groups such as Professor Shiro Kobayashi of Kyoto University, Japan and Professor Klibanov at MIT, USA. Enzyme-catalyzed polymerization here (Enzymatic Polymerization) refers to the "in vitro polymerization by non-biosynthetic pathway" using an enzyme as a catalyst. In recent years, the need for precisely controlling the structure of the functional polymeric material increased tremendously high. Polymerization using an enzyme have been noted as a new method of synthesis of polymers that meet these demand.

Over 5000 different enzyme are known to date, and this number is showing a tendency to increase year by year. Many of these enzymes have been classified into 6 groups as shown in Table 12.3 and enzymes belonging to 3 groups have been reported to be used as polymerization catalyst. Information such as the modification and hydrolysis of the polymer by the respective enzymes are shown in Table 12.3. Difficult synthesis of polymer through conventional synthetic method has been easily conducted by the catalytic action of the enzyme under in vitro condition.

In addition, the enzyme catalyst can be regarded as environmental-friendly catalyst since even the molecular weight of resulting polymer can be increased under mild conditions due to energy-saving properties. The toxicity level caused by the enzyme reaction system is known much less compared with the use of the conventional toxic metal catalyst. In many cases the starting materials of natural

Table 12.3 Enzymes for polymer synthesis (Kobayashi *et al.* 2006)

Enzymes	Polymer syntheses	Polymer modification	Polymer hydrolysis	Typical polymers
Oxidoreductases (peroxidase, laccase, etc.)	○	○	○	Polyphenols, polyanilines, vinyl polymers
Transferases	○	○		Polyesters, polysaccharides
Hydrolases (lipase, protease, cellulase, etc.)	○	○	○	Polyesters, polycarbonates, poly(amino acid)s, polysaccharides
Lyases			○	
Isomerases		○		
Ligases				
Whole-cell	○	○	○	

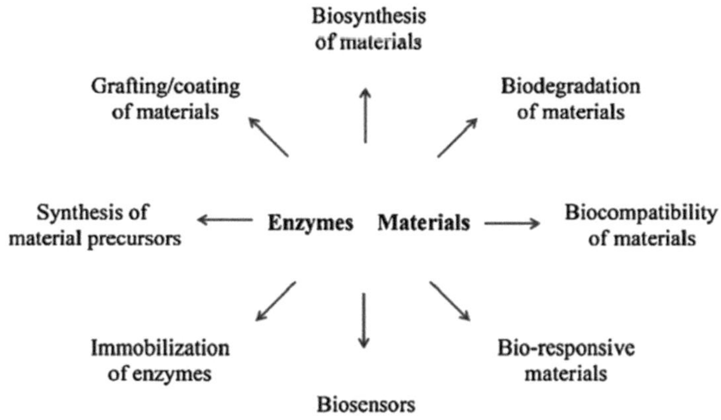

Fig. 12.7 Fields of enzyme application for materials

resources such as biodegradable as well as environment-friendly polymer materials can be utilized to produce polymer material through enzymatic catalysis. That is why enzyme is expected also as a green chemical catalyst in the polymerization because the recirculation of material obtained from the natural world is accelerated. Materials can be modified or changed using enzymes in many aspects, but the common language of enzymes and materials is chemistry (Richter *et al.* 2015). Biomaterials like polyhydroxyalkanoates, cellulose, and lignin are directly and easily related to enzymes as enzymes are involved in their biosynthesis and modification. The enzyme applications related to many fields of materials science are shown in Fig. 12.7. In this chapter, recent trends utilizing the redox enzymes for phenolic polymer and hydrolyzing enzyme (lipases) for ester-based polymer are introduced.

12.2.1 Synthesis of Phenolic Polymers

The formation of a polymer by the oxidation of phenols occurs in the natural polymer such as lignin and melanin through the catalysis of enzyme. Since the mid-1980s, the horseradish peroxidase (HRP) has been used to polymerize the variety of phenols. One of these pioneering works was from Dordick *et al.* (1987) from MIT. Even though redox enzymes such as HRP can effectively generate reactive free radical from phenolics substrate, it is almost impossible to control the coupling reaction between free radicals of phenolics, which is thought as major defect of HRP catalyzed polymerization. However, the enzyme catalytic polymerization of phenol has been attracting attention as environmentally friendly synthesis, which can take the place of conventional phenol formaldehyde resin (novolac resin, resol resin) synthesized with highly toxic formaldehyde.

Peroxidase catalyzes the oxidation of various phenolic substances with hydrogen peroxide as an electron acceptor.

This enzyme, peroxidase, is present in a wide range of plant cells, and is deduced to be involved in defense against reactive oxygen species in the body alive. Polymerization mechanism by such peroxidase has long been known and peroxidases such as HRP and soybean peroxidase (SBP) are typical heme protein containing iron porphyrin at the reaction active sites for the oxidation of the substrate and the release of a water molecule. As in Fig. 12.8, peroxidase can form two kinds of active oxygen complex by elimination of a hydrogen atom from the substrate as a result oxidizes the substrate by using the peroxide as the oxidizing agent. For the phenol oxidation by HRP, radical called phenoxyl radical is generated and polymerization is carried out by coupling of these radicals. For a simple phenol containing no substituent at phenol ring, soluble and structure-controlled polymer cannot be easily obtained since at least four reactive sites at one monomer phenol radical are present during the oxidative coupling.

The polymerization by using HRP as catalyst in the reaction medium consisting of mixed solvent of methanol and soluble buffer resulted in the soluble polyphenolics in a high yield. The supposed structure of polyphenolics was composed of phenylene and oxyphenylene bond as shown in Fig. 12.9.

For peroxidase catalyzed polymerization of phenolics, aqueous organic solvents are being commonly used as reaction medium. Prior to polymerization, the monomers phenol and the enzyme are dissolved in a homogeneous reaction medium, but the polymerization initiated by the action of hydrogen peroxide (oxidizing agent) results in the reduced solubility of the resulting polymer. In accordance with the decreased solubility of the reaction system the resulting polymer can be recovered through filtration or centrifugal precipitation.

Substituted phenols with an alkyl group in the *meta* position can be polymerized as soluble polymer as shown in Fig. 12.9. Poly (m-cresol) obtained in aqueous methanol solution (pH 7, phosphate buffer) was soluble in the polar solvent and the glass transition temperature was at least 200 °C and the polymer exhibited biodegradation properties in the soil. The smaller the size of the alkyl group of monomer phenolics, the more increased yield of polymer for HRP catalyst. However, this trend seemed opposite for soybean peroxidase (SBP) with increased the yield of polymer from the larger the size of substituent group. Since polymerization of phenol proceeds by free radical mechanism, the control of polymerization is not easy, however, changing the solvent composition (mixing ratio) of methanol and phosphate buffer can result in the differential range of phenylene units from 40 to 70%, which implies that the structure control is possible at certain degree. By varying log P of organic solvent phenylene/oxyphenylene unit ratio was controlled in the range of 94/6–4/96. In addition, polyethylene glycol (PEG) as a template made oxidative polymerization of the phenol conducted in pure water solution and resulting polymer containing 90% or more phenylene units was synthesized.

Recently, it was reported the polymerization of the phenol with laccase in the oxidation–reduction as a catalyst was performed. For peroxidase, hydrogen

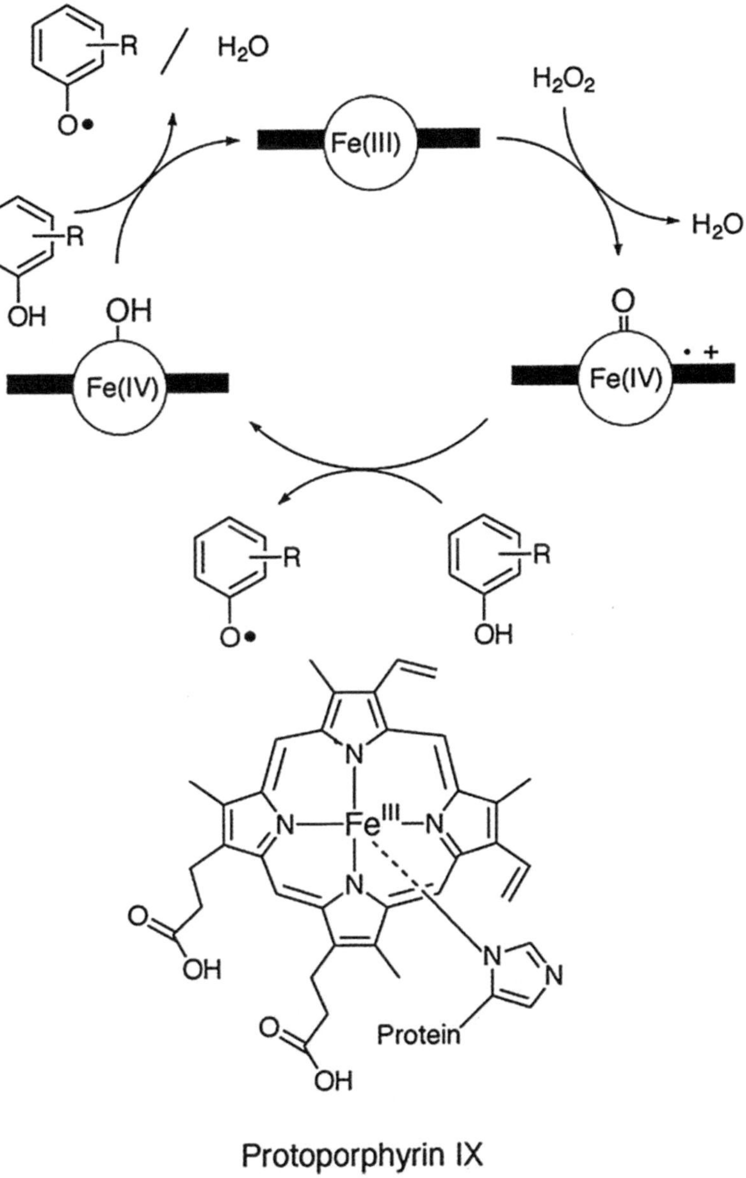

Fig. 12.8 The mechanism of phenol substrate oxidation at the active center of peroxidase (Kobayashi *et al.* 2006)

peroxide is necessary in the polymerization as the oxidizing agent, but when using a laccase has the advantage that the oxygen in the air can be directly used an oxidizing agent. Polymerization behavior is very similar to the case of using

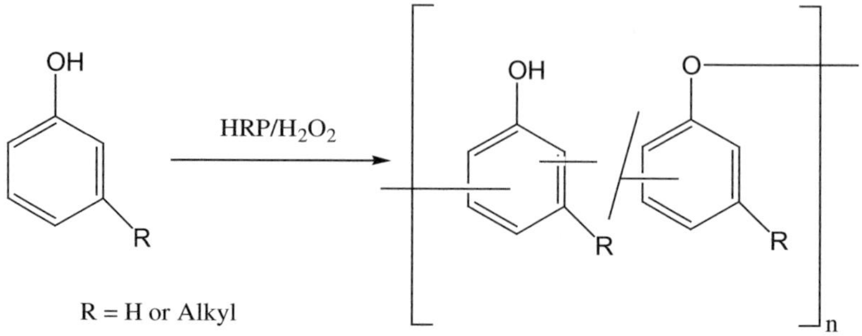

Fig. 12.9 Polymerization of phenolics through oxidative coupling (Akkara *et al.* 1991)

peroxidase as a catalyst, and reported to be also possible precision control of the structure by the solvent composition.

A summary of the characteristics of the polyphenolic synthesis using an enzyme as a catalyst as follows: (1) do not use a strong toxic formaldehyde, (2) is very high catalytic activity (turnover number over 100,000), (3) phenolic compounds of the wide range of structures is polymerized otherwise difficult to yield with conventional polymerization method, (4) it is possible to control the solubility and molecular weight by the reaction conditions, (5) polymerization proceeds under mild conditions (room temperature, normal pressure), (6) the operation is simple, relatively easy separation and purification. Such a strategy of choice for many benefits reported on the synthesis of a variety of phenol-based polymer using the enzyme has been released.

Regioselective Polymerization

As shown in Fig. 12.10a, enzyme can make it possible to synthesize a phenolic polymer which is selectively oxidized otherwise cannot be obtained by using a different metal catalyst. Using a general oxidation catalyst results in the most radical generation and crosslinkage formation on acrylic groups, on the while for HRP/hydrogen peroxide system selectively phenoxy radicals are generated to form C–C bond or C–O bond. In the case of arbutin being used as a core material of whitening cosmetics can only found on the *ortho* position that as a result radical coupling regioselective synthesis of a polymer as shown in Fig. 12.10b. In addition, as shown in Fig. 10.3c are a case of syringic acid, a linear polymer is synthesized in this reaction while carbon dioxide is released as leaving group, which cannot be carried out by typical metal catalyst. Other method for synthesizing a phenolic polymer in addition to radical generation is using halogen-substituted aromatic compounds derivatized through nucleophilic substitution reaction using metal phenolate as catalyst. However, these methods employ halogenated monomers environmentally in question and at the same time there is a problem it is necessary to remove the salts formed with high temperature of the reaction.

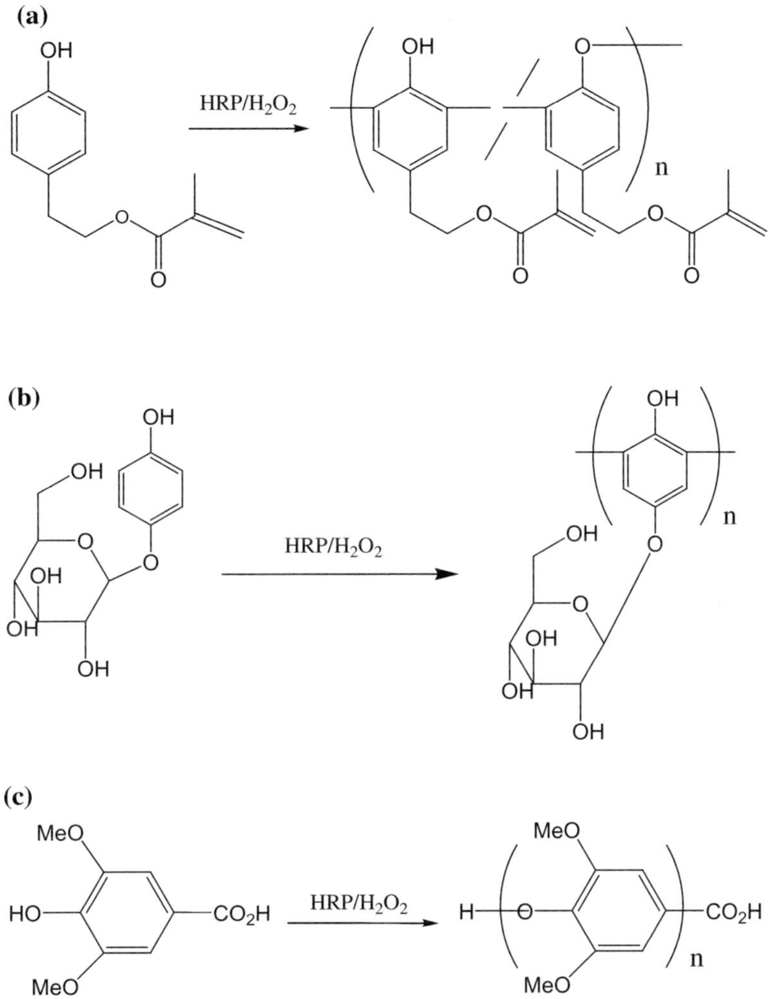

Fig. 12.10 Examples of regioselective polymerization of phenolics by enzymes (Ikeda *et al.* 1998)

Chemoselective Polymerization **Using Enzyme**

If the phenolics harbor functional groups at their side position, chemoselective reaction between phenolic groups and functional groups can be achieved by peroxidase. In the case of phenol having an ethynyl when polymerized using the HRP/hydrogen peroxide it can be synthesized polymer in a state retained diacetylene group as compared with the other metal catalysis (copper/diamine complex).

It was reported that 4-phenylazophenol was synthesized into polymer a number average molecular weight of 3000 by HRP. As can be seen in Fig. 12.11, resulting polymers are mainly known to have the C–C coupling formed in the *ortho*

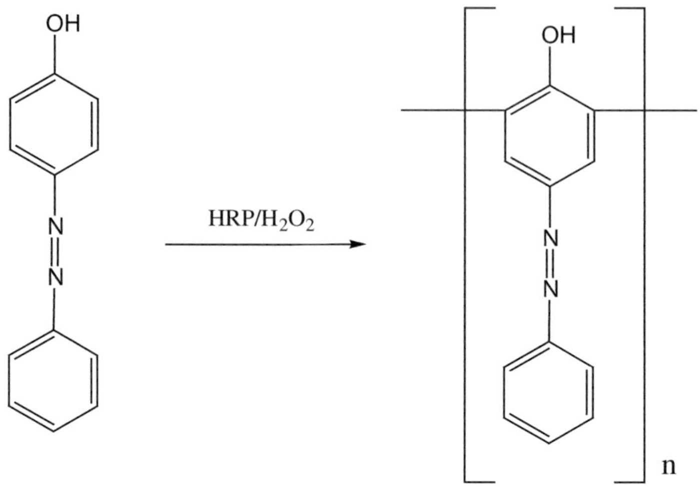

Fig. 12.11 Chemo- and regioselective polymerization of phenol containing azo functional group by peroxidase (Alva *et al.* 1998)

position. It is noted that this polymer showed a fluorescent light and a number of unique optical properties.

Meta substituted natural phenol harboring unsaturated double bonds, CNSL (Cashew Nut Shell Liquid) is being produced about 1 million tons per year in Brazil and India as a sort of by-products after a cashew nut production. Only about 20,000 tons of CNSL consisted of three derivatives (Fig. 12.12) are known to be used in industrial applications. The CNSL and formaldehyde or hexamethylenetetramine has been used in many industrial applications to form a copolymer with of other phenols. After cardanol, one of pure constituent of CNSL was separated and purified through distillation, enzymatic polymerization was carried out.

Peroxidase may form a polymer in a state that does not impair the unsaturated double bonds present in the *meta* position, which is another example of chemoselective polymerization by enzyme as shown in Fig. 12.12. The polymer reaction is greatly influenced by the environment present in organic solvent.

In general, the enzyme reaction in an organic solvent generally depends on log *P* values, but it appears that the polymerization of cardanol was carried out with high yield in 2-propanol whose log *P* value is considerably small polar solvent. Average molecular weight of polycardanol polymerized using the SBP was shown to have from 8000 to 12,000 through GPC measurement, the average yield was determined to be 70% or more as shown in Table 12.4.

Figure 12.13 shows the comparison of the experimental results about polycardanol stain resistance (Anti-fouling) toward the microorganism after the hard film coating using a cobalt-based catalyst on slide glass surface. Targeted microorganism has been known as a biofilm forming bacteria *Pseudomonas fluorescens*.

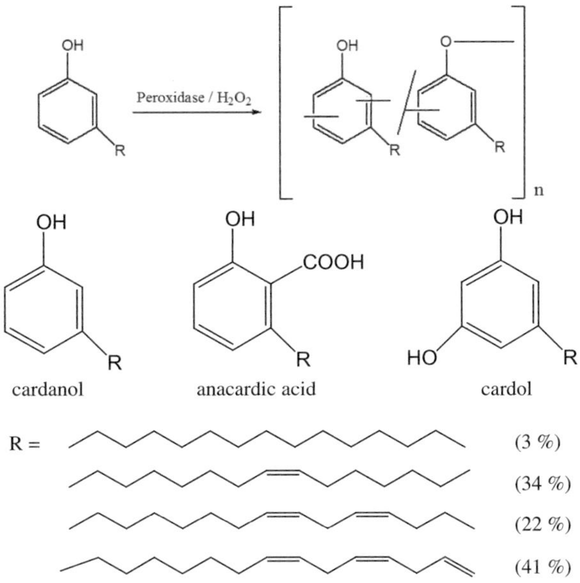

Fig. 12.12 Chemical structures of constituents of CNSL and chemoselective polymerization of cardanol (Kim *et al.* 2005)

Table 12.4 Polymerization of cardanol in various aqueous organic solvents (Kim *et al.* 2003)

Cardanol (m mol)	Organic solvent	Enzymes	Yield (%)	Mn	Mw
2.0	Methanol	SBP	42.5	3540	12,808
2.0	Ethanol	SBP	50.4	4096	10,974
2.0	Isopropanol	SBP	72.5	3411	8221
2.0	*t*-butanol	SBP	0	–	–
2.0	1,4-dioxane	SBP	0	–	–
2.0	Isopropanol	HRP	0	–	–

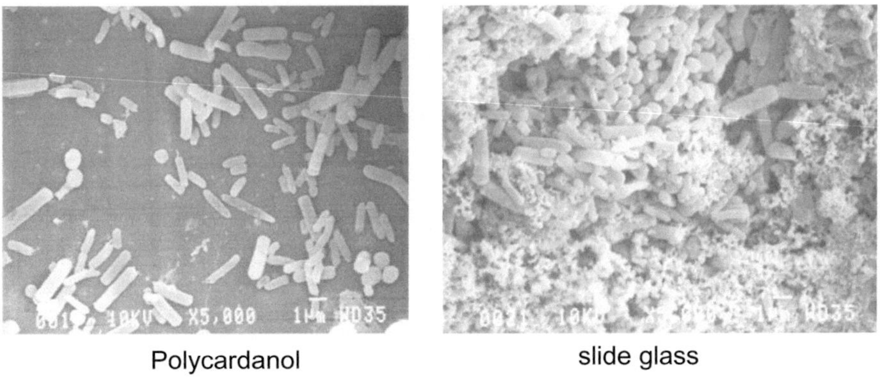

Fig. 12.13 SEM observation of biofilm on polycardanol coating and conventional slide glass (Kim *et al.* 2003)

These microorganisms on the cured coating film of polycardanol were incubated in culture plate and then put together to observe the shape of the biofilm.

Urushiol is a monomer from *Rhus vernicifua* tree and have been used for coating wood materials from long time ago in Orient. Since Urushiol structure is very similar to cadanol except one more –OH at the ortho position, polymerization of Urushiol is possible by laccase to make oriental lacquer (Kim and Yoo 2001).

12.2.2 Synthesis of Polyester

Aliphatic polyesters have characteristics of excellent biodegradability. In addition, biocompatible and permeable properties make aliphatic polyesters acceptable for biomedical applications. For example, resorbent implant may be the first candidate in tissue engineering. Recently, enzymes such as lipases and esterases have been increasingly used as biocatalysts for the production of aliphatic polyesters to avoid the trace metallic residues which can be found in using chemical catalysts (Kobayashi and Makino 2009; Zhang *et al.* 2014). Enzymatic polymerization has been regarded as environment-friendly for polymeric materials, providing a good example of "green polymer chemistry". Lipase/esterase-catalyzed synthesis of aliphatic polyesters proceeds via two major modes (Fig. 12.14): (1) ring-opening polymerization of lactones, (2) polycondensation, and self-polycondensation of oxyacids or their esters subunit.

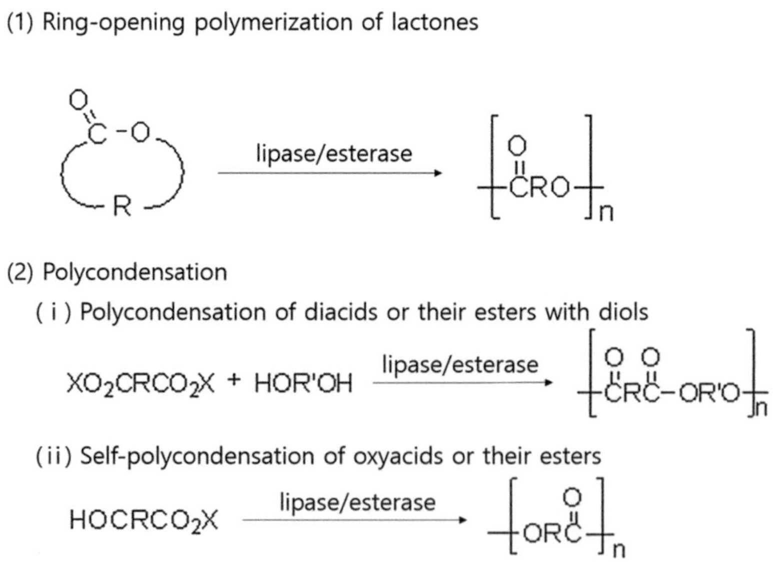

Fig. 12.14 Two major modes of lipase/esterase-catalyzed polyester synthesis

In enzymatic ring-opening polymerization, no water or alcohol are produced in contrast to polycondensation. Thus, enzymatic ring-opening polymerization has been extensively investigated to synthesize polyesters, polycarbonates, polyphosphates and polythioesters using cyclic monomers as substrates (Yang *et al.* 2011). In enzymatic polycondensation reaction, various dicarboxylic acids and their activated or nonactivated esters polymerized with diols, and mercaptoacids or their esters have been employed to construct polyesters (Yu *et al.* 2012).

Though enzymatic polymerization has been rapidly developed, the range of available polymers is still limited. This could be overcome by a combination with chemical methods (Yang *et al.* 2014). To date, enzymatic polymerization has been successfully combined with atom transfer radical polymerization, kinetic resolution, reversible addition–fragmentation chain transfer, click reaction, carbene chemistry and ring-opening polymerization to construct polymeric materials like block, brush, comb and graft copolymers, hyperbranched and chiral polymers (Xiang *et al.* 2014). Thus, combining enzymatic polymerization with chemical methods will be an efficient tool to produce many polymer materials.

Enzyme-Catalyzed Synthesis of Polyester

Compared with conventional enzymes optimized at low temperature, thermophilic enzymes tolerant at high temperature originated from thermophiles have been recognized as potential catalysts in various applications due to their superior stability against organic solvents, high temperature and chemical denaturants. Feng's group once investigated the polyester synthesis using thermophilic esterase from *Archaeoglobus fulgidus* and a thermophilic lipase from *Fervidobacterium nodosum*, using ring-opening polymerization of ε-caprolactone as a model (Ma *et al.* 2009; Li *et al.* 2011a, b). By employing these enzymes, complete conversion of monomers to polyesters in organic solvents was achieved, yielding oligoesters with Mn <2500 g/mol. It was also found that these two enzymes had stronger affinity toward the monomer ε-caprolactone, which was confirmed by enzyme kinetic analysis and through docking simulation between the enzyme and the ligand.

Through a ring-opening polymerization of lactone compounds as a starting material synthesis of the polyester polymers is catalyzed by enzyme lipases. Historically, the first one was reported by Professor Kobayashi group in Kyoto University of Japan (Kobayashi and Makino 2009) for polymer synthesis via ring-opening reaction of the lactone compound. Lactone compounds consisting of 16 carbons can be polymerized through a ring-opening reaction wherein produced polymer properties are known to be affected by the characteristics of the enzyme catalyst and lactone of the starting material. Of course, when using the ring-opening polymerization of the lactone, even though general chemical catalyst is also available as a catalyst, particularly the lactone ring compound having a large size can be more easily polymerized by using enzymes as shown in Fig. 12.15.

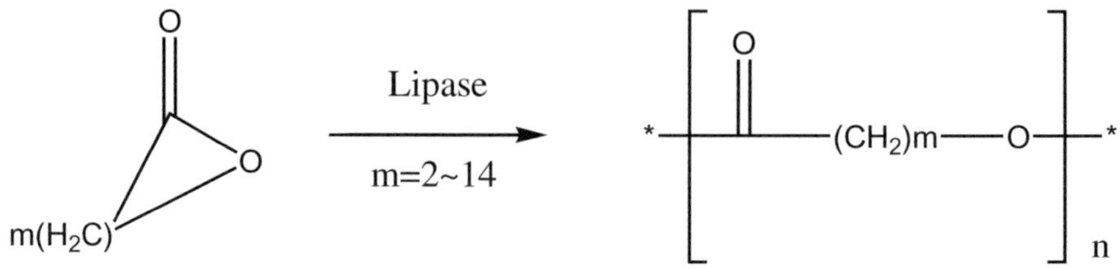

m=2; PL	m=2; poly(3HP)
m=3; BL	m=3; poly(4HB)
m=4; VL	m=4; poly(5HV)
m=5; CL	m=5; PCL
m=10; UDL	m=10; poly(UDL)
m=11; DDL	m=11; poly(DDL)
m=14; PDL	m=14; poly(PDL)

Fig. 12.15 Ring-opening polymerization of lactone compounds catalyzed by lipase enzyme (Kobayashi *et al.* 2006) *PL* propiolactone, *Poly(3HP)* poly-2-hydroxypropionate, *BL* butyrolactone, *Poly(4HB)* poly-3-hydroxybutyrate, *VL* valerolactone, *Poly(5HV)* poly-4-hydroxyvalerate, *CL* caprolactone, *PCL* polycaprolactone, *UDL* 11-undecanolide, *DDL* 12-dodecanolide, *PDL* 15-pentadecanolide

12.2.3 More Cases

Many other cases have been introduced. Homopolymers as well as copolymers are also important for industrial applications. To meet the property demand from the customers, many different and diversified properties are required. For this purpose, copolymer and polymer blend can be a solution. Modification of the natural polymers such as starch and cellulose is another way to change the properties of the polymer. Even though many polymers are being synthesized using chemical means, some polymers can be also synthesized using enzyme as catalyst. In this case, properties of the enzyme-catalyzed polymers are different from the chemically synthesized polymers, in molecular weight, biodegradability, etc. Deplolymerization of the polymer is also an important issue—such as hydrolysis of starch, cellulose, and hemicellulose. Current issues are hydrolysis of polysaccharides and bioremediation of the polymers in nature environment.

Further Discussion

1. Cephalosporins are very useful antibiotics. What progress have been made recently to improve the technology for cephalosporins, and what technology are to be developed for further development?
2. Cytochrome p-450 enzymes are widely investigated. What reactions and applications are possible with p-450 enzyme?

3. Nowadays, chemicals production from biomass or using biotechnology is very promising. List up examples which are being commercialized or near commercialization and discuss on the technology.
4. List up and discuss the most important technology for the bio-based chemicals production to replace the petrochemicals.
5. Cellulose and polysaccharides are also polymers from nature. Can we synthesize cellulose-like polymers by enzymatic means?
6. Many polyesters by enzymatic means are biodegradable. What characteristics of polyesters make the polyester biodegradable? Molecular weight of polyester is very important for commercial applications of polyesters. How can we control the molecular weight of polyesters when using enzymes?

References

Akkara JA, Seneccal KJ and Kapaln DL. Synthesis and characterization of polymers produced by horseradish peroxidase in dioxane. *J. Polym. Sci. A Polym. Chem.*, 1991, 29:1561–1574.

Alva KS, Lee TS, Kumar J and Tripathy SK. Enzymatically synthesized photodynamic polyaniline containing azobenzene groups. *Chem. Mater.*, 1998, 10:1270–1275.

Chandel AK, Rao LV, Narasu ML and Singh OV. The realm of penicillin G acylase in β-lactam antibiotics. *Enzyme Microb Technol*, 2008, 42:199–207.

Dordick JS, Marletta MA and Klibanov AM. Polymerization of phenols catalyzed by peroxidase in nonaqueous media. *Biotechnology and Bioengineering*. 1987, 30:31–36.

Drauz K and Waldmann H. ed, *Enzyme Catalysis in organic synthesis; a comprehensive handbook. Volume I, II*. VCH, 1995.

Hanai T, Atsumi S. and Liao JC. Engineered synthetic pathway for isopropanol production in *Escherichia coli*. *Appl. Environmental Microbiology,* 2007, 73:7814–7818.

Ikeda R, Sugihara J, Uyama H and Kobayashi S, Enzymatic oxidative polymerization of 4-hydroxybenzoic acid derivatives to poly(phenylene oxide)s. *Polym Int.*, 1998, 47:295.

Kim JW and Yoo YJ. A new detergentless micro-emulsion system using uroshiol as an enzyme reactor system. *J. Microbio. Biotechnol.* 2001, 11:369–375.

Kim YH, Won KH, Kwon JM, Jeong HS, Park SY, An ES and Song BK. Synthesis of polycardanol from renewable source using a fungal peroxidase *Coprinus cinereus*, *J. Molecular Catalyst B. Enzymatic*, 2005, 34:33–38.

Kim YH, An ES, Song BK, Kim DS and Rahul Chelikani, Polymerization of cardanol using soybean peroxidase and its potential application as anti-biofilm coating material, *Biotechnology Letters*, 2003, 25:1521–1524.

Kobayashi S, Ritter H and Kaplan D. *Enzyme-catalyzed synthesis of polymers*, Springer, 2006.

Kobayashi S and Makino A. Enzymatic polymer synthesis: an opportunity for green polymer chemistry. *Chem. Rev,* 2009, 109:5288–5353.

Liu H and Lu T. Autonomous production of 1,4-butandiol via a *de novo* biosynthesis pathway in engineered *Escherichia coli*. *Metabolic Engineering*, 2015, 29:135–141.

Li Q, Li G, Yu S, Zhang Z, Ma F and Feng Y. Ring-opening polymerization of ε-caprolactone catalyzed by a novelthermophilic lipase from *Fervidobacterium nodosum*. *Process Biochem.*, 2011a, 46:253–257.

Li Q, Li G, Ma F, Zhang Z, Zheng B and Feng Y. Highly efficient ring-opening polymerization of ε-caprolactone catalyzed by are combinant *Escherichia coli* whole-cell biocatalyst. *Process Biochem.,* 2011b, 46:477–481.

Ma J, Li Q, Song B, Liu D, Zheng B, Zhang Z and Feng Y. Ring-opening polymerization of ε-caprolactone catalyzed by a novelthermophilic esterase from the archaeon *Archaeoglobus fulgidus*. *J. Mol. Catal. B: Enzyme*, 2009, 56:151–157.

Parmar A, Kumar H, Marwaha S and Kennedy JF. Advances in enzymatic transformation of penicillins to 6-aminopenicillanic acid (6-APA). *Biotechnol Adv*, 2000, 18:289–301.

Richter M, Schulenburg C, Jankowska D, Heck T and Faccio G. Novel materials through nature's catalysts. *Mater. Today,* 2015, doi:10.1016/j.mattod.2015.04.002.

Schulze B and Wubbolts MG. Biocatalysis for industrial production of fine chemicals. *Current Opinion in Biotechnology*, 1999, 10:609–615.

Thomas SM, Di Cosimo R and Nagarajan V. Biocatalysis: applications and potentials form the chemical industry, *Trends Biotechnol.* 2002, 20:238–242.

Wu Q, Chen CX, Du LL and Lin XF. Enzymatic synthesis of amoxicillin via a one-pot enzymatic hydrolysis and condensation cascade process in the presence of organic co-solvents. *Appl Biochem Biotechnol,* 2010, 160:2026–35.

Xiang S, Zhang Q, Zhang G, Jiang W, Wang Y, Zhou H, Li Q and Tang J. Facile synthesis of block copolymers by tandem ROMP and eROP from esters precursors. *Biomacromolecules*, 2014, 15:3112–3118.

Yamada H. and Nagawawa T. Production of useful amides by enzymatic hydration of nitriles. *Ann. N.Y. Acad. Sci.*, 1990, 613:142–154.

Yang Y, Yu Y, Zhang Y, Liu C, Shi W and Li Q. Lipase/esterase-catalyzed ring-opening polymerization: a green polyester synthesis technique. *Process Biochem.,* 2011, 46:1900–1908.

Yang Y, Zhang J, Wu D, Xing Z, Zhou Y, Shi W and Li Q. Chemoenzymatic synthesis of polymeric materials using lipases as catalysts: A review. *Biotechnol. Adv.*, 2014, 32: 642–651.

Yu Y, Wu D, Liu C, Zhao Z, Yang Y and Li Q. Lipase/esterase-catalyzed synthesis of aliphatic polyesters *via* polycondensation: a review. *Process Biochem.,* 2012, 47:1207–1236.

Zhang J, Shi H, Wu D, Xing Z, Zhang A, Yang Y and Li Q. Recent developments in lipase-catalyzed synthesis of polymeric materials. *Process Biochem.*, 2014, 49:797–806.

Chapter 13
Enzymes for Food and Energy

13.1 Enzymes for Food Industry

Enzymes exist very widely in natural and processed foods. Industrial production of enzyme used for food processing since 1874, when Christian Hansen, a Danish scientist, obtained rennin (a kind of protease) from calves' stomachs to apply for cheese preparation (Nielsen *et al.* 1994). Currently, many enzymes commonly consumed in food industry are originated from recombinant microbials. Often native enzymes cannot meet the demand of industrial scale, genetic techniques and enzyme engineering thus provide enzymes with improved properties for their development and manufacture.

Enzyme preparation used in food processing contains substances such as viscosity modifiers, preservatives, and dispersants. Other enzymes and metabolites originated from microorganisms or substrates used to culture microorganisms can be contained. However, all these materials are known to be suitable for food and are expected to meet the requirement of current good manufacturing practice (cGMP). Safety considerations are required for both native and recombinant microorganism-derived enzymes. The essential procedure to guarantee the safety of enzyme is to use safe microbial as host (Pariza and Johnson 2001). Food and Drug Administration (FDA) of each country reviews generally recognized as safe (GRAS) affirmation petitions for enzyme preparation.

13.1.1 Dairy Products

Dairy foods manufacturing industry uses many enzymes. Rennet preparations containing acid proteases extracted from animal tissue is the well-known dairy enzyme preparation (Harboe and Budtz 1999). Rennin is the active enzyme

Y.J. Yoo *et al.*, *Fundamentals of Enzyme Engineering*,
DOI 10.1007/978-94-024-1026-6_13

Table 13.1 Enzymes applied in dairy food industry

Enzymes	Application examples
Acid proteinases	Milk coagulation
Neutral proteinases and peptidases	Accelerated cheese ripening, debittering, enzyme-modified cheese, production of hypoallergenic milk-based foods
Lipases	Accelerated cheese ripening, enzyme-modified cheese, flavor-modified cheese, structurally modified milk fat products
β-Galactosidase	Lactose-reduced whey products
Lactoperoxidase	Cold sterilization of milk, milk replacers for calves
Lysozyme	Nitrate replacer for washed-curd cheeses and cheeses with eyes (e.g. Emmental)

found in rennet. It removes a highly charged peptide segment from κ-casein. Destabilized casein micelles make aggregation in the form of milk clot, then are acidified by lactic cultures, and finally cheese curd is formed. The dairy foods sector also use other enzymes, see Table 13.1.

Proteinases, peptidase, and lipases are the enzyme classes used in the commercial ripening technology for cheese manufacturing. This list may extend if current researches on metabolic enzymes such as acetyl-CoA synthases and amino acid-catabolizing enzymes to generate volatile esters and sulfur compounds would be successful. The mixture of more than one class of enzyme enhances or accelerates the cheese maturation process.

Accelase® is one of the commercially available enzyme preparations in the production of cheeses and reduced fat variants. Food-grade proteinases, aminopeptidases, esterases, and flavor enzymes found in lactic acid bacteria are used in the formulation. This enzyme preparation can reduce the cheese maturity period from 9 to 5 months. Additionally, amino acid pool in the cheese can be increased, flavor and aroma are enriched and bitterness is suppressed. Another commercially available enzyme preparation for cheese-ripening are Rulactine (Rhone-Poulenc) and Flavorage (Chr. Hansen, US, Inc.), which are a proteinase from *Micrococcus* sp. and a lipase from *Aspergillus* sp. with proteolytic enzymes respectively. However, these two enzymes preparations have little information as to their efficacy in cheese manufacture.

Rennet. Calf rennet is considered to be a milk-clotting enzyme optimal for cheese making. Calf rennet is typically 80–90% rennin (EC 3.4.23.4). Pepsin (EC 3.4.23.1), the second enzyme in calf rennet, is thought to help the ripening in maturing cheese but without a concrete proof. Regions with a shortage of calves applied modern production methods to extract the important enzymes from yeast molds and fungi. Cheese makers nowadays make excellent cheese quality using pure rennin from genetically modified yeast and fungi expressing calf (pro)chymosin gene. Rennet preparations can also be obtained from sheep, goat and pig, but not functionally ideal for cow milk (Foltmann 1992).

Lysozyme. In Gouda, Emmental and other hard and semi-hard cheeses, defective textures and irregular holes caused by butyric fermentation can occur. This can be controlled by lysozyme (EC 3.2.1.17), which suppresses spores and vegetative cells. Since it binds to cheese curd, it is stable in the cheese matrix for long periods and less enzyme is lost on whey separation.

Lipases. Lipases intensify flavor development in cheese. Lipolysis of cheese not only contributes to the added flavor but it also accelerates ripening of the cheese. On the other hand, lipases released short to medium chain fatty acids and their chemical derivatives from milk for the creamy, buttery, and cheesy aroma of lipolyzed milk fat (LMF).

Transglutaminases. Protein-glutamine γ-glutamyltransferase (EC 2.3.2.13) is efficient in reducing syneresis in acid milk gels. It has been investigated for the texture improvement and shelf life of yogurt (Motoki and Seguro 1998). Its application to gelation of caseins and whey proteins has been considered but the technology is not widespread yet.

13.1.2 Bread Making

The application of enzymes is a huge contributor to the improvement in quality in terms of flavor, texture, and shelf life of bakery products. Baking enzymes are primarily used as flour additives and dough conditioners. Baking industry makes use of amylase, protease, xylanase, oxidase and lipase.

Hydrolyases

Amylases. α-amylases (EC 3.2.1.1) are endoglucanases that hydrolyze α-1,4 and α-1,6 linkages of starch randomly. Suitable amount of amylase can lead to high quality of dough and final product. However, excessive amounts lead to sticky dough due to possible extensive degradation of starch. Although β-amylase and pullulanase two other amylases are capable of hydrolyzation of starch, their effect on dough properties and bread quality are limited.

Xylanases. Xylanase converts water-insoluble hemicelluloses into soluble polysaccharides. These dissolved hemicelluloses in water have capability to bind water in the dough, which result in the decrease of dough firmness and increase of volume. Xylanases make the dough more suitable for mass production through machine because it has low tendency to stick to machine surface (Rouau 1993). In bread making, the activity of xylanase is unpredictable, thus bread manufacturers and bakeries need to find the optimal amount and mix of xylanases by trial for each application.

Lipases. Lipases can be employed as emulsifiers since the lipases hydrolyze polar wheat lipids to generate emulsifying lipids in situ (Collar *et al.* 2000). 1,3-specific lipases can improve dough-handling properties, dough strength and stability and dough machinability, and increase oven spring. Lipase enriches the flavor of breads by releasing short-chain fatty acids from lipid. Aside from the above benefits, the service-life of most of bakery products can be extended by the help of lipases.

Proteases. During bakery production, a protease originated from fungi is used to alter the properties of the mixture containing high gluten content. When these enzymes are mixed in the blend, the mixture experiences partly hydrolysis, which helps to increase softness. Proteases are added to decrease necessary mixing time, to maintain dough conformity, to control gluten firmness in bread, to modify bread texture and to enhance the flavor. Proteases give great effects on dough rheology and the quality of bakery products owing to influence on the gluten texture.

Oxidoreductases

Lipoxygenases. Linoleate oxygen oxidoreductase (EC 1.13.11.12) is a nonheme iron-containing dioxygenase. This enzyme catalyzes the oxidation of polyunsaturated fatty acids containing *cis,cis*-1,4-pentadiene system to form fatty acid hydroperoxides using oxygen as oxidizing reagent. The commercially available lipoxygenases comes from soybean flour and other beans (e.g., fava beans). The short live and reactive radicals generated during enzyme catalyzed oxidation can oxidize pigments and protein thiol groups contained in the dough, which results in the production of hydroxyacids (Boussard *et al.* 2012). In bread making, lipoxygenase from soya flour is applied as bleaching agent to make the crumb white, as well as being used to improve the dough rheology, tolerate mixing, increase the volume of the loaf, and stabilize the gluten. However, lipoxygenase activity can bring about unwanted flavors such as ketodienes.

Glucose oxidases. β-D-glucose:oxygen:1-oxidoreductase (EC 1.1.3.4) oxidizes β-D-glucose to D-glucono-δ-lactone and hydrogen peroxide simultaneously. Glucose oxidase depletes leftover glucose and oxygen in foods to extend the expiration date. This enzyme can be employed as oxidizing catalyst instead of potassium bromated, which is recognized as carcinogenic, in bread making. The hydrogen peroxide from the enzymatic reaction has antimicrobial properties. On the other hand, hydrogen peroxide promotes the oxidative crosslinks between either cysteines or tyrosines in the gluten structure. More disulfide cross-linking and/or oxidative gelation on the gluten network endows dough higher chance of machine application, improved gas retention, which result in increased bread volume and fine crumbs (Bekes *et al.* 1994).

Laccases. Benzenediol oxygen oxidoreductase (EC 1.10.3.2) belongs to multi-copper containing polyphenol oxidases at the reactive center. The laccase added to dough in baking helps to elevate strength of gluten matrix through oxidation. It also added for the machinability of dough was intensively increased due to reinforced strength and longer duration time along with diminished sticky property. A number of industrial laccases are available but almost no laccase is commercialized directly for baking.

13.1.3 Starch Processing

Starch, a renewable source, is an industrial raw material. It can be widely found as storage compound in plant materials such as corn, tapioca, wheat, rice, and potato.

Amylose and amylopectin are major components of starch molecules. Amylose is linear polymerized product composed of glucose as subunit via α-1,4 glycosidic linkages. On the while amylopectin is partly branched polymer up to 5% α-1,6 linked side chains (van der Maarel *et al.* 2002). Depending on the sources, starches containing various ratios of amylose and amylopectin are found in starch granules as packed state. These components of starch can be altered through various chemical and enzymatic processes depending on the demand of food ingredients.

There is a list of enzymes reaction on starch molecules (see Fig. 11.1). Since starch polymer is a complex structure, combination of different hydrolases specific to various bonds in starch are required. The common enzyme for starch hydrolysis is amylases. Amylases can be subgrouped into four groups: endoamylases, exoamylases, debranching enzymes, and transferases (van der Maarel *et al.* 2002; Nigam and Singh 1995).

Endoamylases. Endo-acting enzymes hydrolyze internal α-1,4 bonds in the amylose or amylopectin chain resulting in α-anomeric products. Enzyme α-amylase (EC 3.2.1.1) depolymerizes starch by the hydrolysis of internal α-1,4-*O*-glycosidic bonds, still maintaining α-anomeric configuration in depolymerized one.

Exoamylases. Exoamylases hydrolyze α-1,4 or α-1,6 bonds of the external glucose residues of amylose or amylopectin and yield α- or β-anomeric products. Glucoamylases (EC 3.2.1.3) and α-glucosidases (EC 3.2.1.20) are among the exoamylase enzymes. β-amylases (EC 3.2.1.2) can exclusively hydrolyze α-1,4 bonds producing maltose and β-limit dextrin.

Debranching enzymes. They hydrolyze α-1,6 bonds of amylopectin and/or glycogen polymer exclusively leaving long linear polysaccharides. Debranching enzymes are subclassified into three distinctive groups depending on their specificity: (1) pullulanases from microorganism and pullulan-6-glucanohydrolases, (2) isoamylases, and (3) amylo-1,6-glucosidases found in higher microorganisms. Pullulanase (EC 3.2.1.41) especially from microbial origin gain attention due to its specific activity on α-1,6 linkages in pullulan. Another debranching enzyme, isoamylase (EC 3.2.1.68), is the only enzyme known to remove branches from glycogen completely (Fig. 13.1).

Transferases. Transferases cut off the α-1,4 glycosidic bond of the donor molecule and transfer a donor fraction to a glycosidic acceptor to form a new glycosidic linkage. Amylomaltases (EC 2.4.1.25) catalyze the transfer to the non-reducing ends of the side chains of amylopectin. Glycogen branching enzymes (EC 2.4.1.18) cleave an α-1,4 glycosidic linkage and form a new α-1,6 linkage after all.

Enzymes in Starch Hydrolysis

Gelatinization, liquefaction and saccharification are the steps involved in the starch hydrolysis (Fig. 13.2). In gelatinization, the water–starch slurry is quickly heated to break open the starch granules causing amylose to leach out. Before steam is injected into the jet-cooker, pH is adjusted for best amylase function, calcium is also added for amylase stability and finally a thermostable amylase is added such as that of *Bacillus licheniformis*. Gelatinization takes place at about 105 °C for about 5 min.

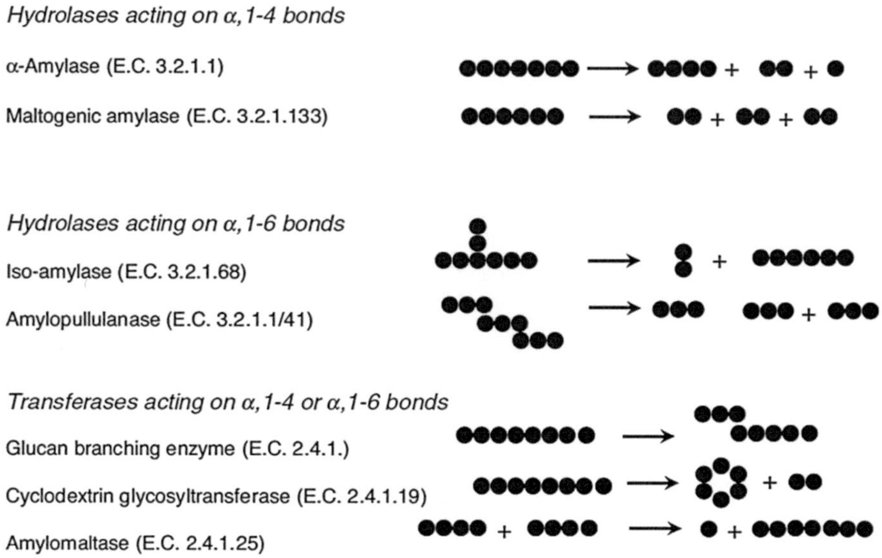

Fig. 13.1 Schematic diagram for the hydrolysis of starch using different enzymes (van der Maarel *et al.* 2002)

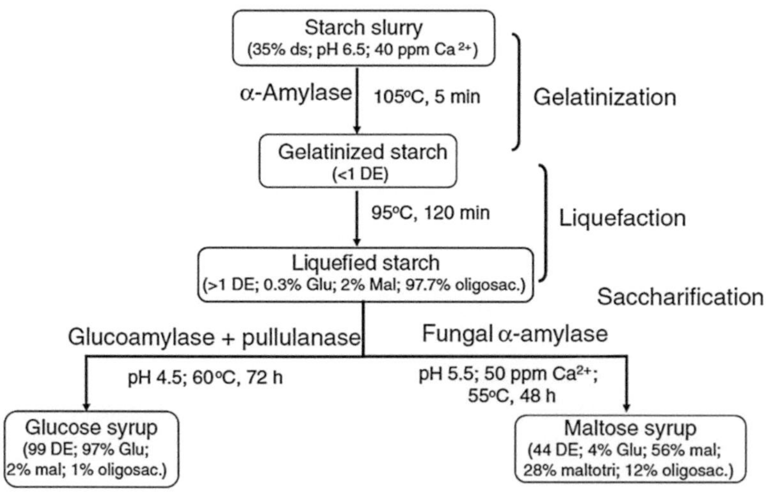

Fig. 13.2 Steps in industrial starch hydrolysis (van der Maarel *et al.* 2002)

The next step to gelatinization is liquefaction. The temperature of the slurry is adjusted to 95–100 °C and hold for 1–2 h. During this process, the thermostable amylase hydrolyze the α-1,4 linkages of the starch molecules drastically reducing the viscosity of the gelatinized starch solution. The depolymerization of starch during liquefaction produces dextrins. Liquefaction is permitted to be carried out until the target dextrose equivalent (DE) value (defined as "the number of reducing ends relative to a pure glucose of the same concentration") is obtained.

In saccharification step, pullulanase, glucoamylase, β-amylase, or an α-amylase are added for further hydrolyzing starch containing target DE value into malto-dextrins, maltose, or glucose syrups, respectively. Saccharification of starch is carried out at 60 °C and pH of 4–5. The temperature after liquefaction is cooled down as quickly as possible to avoid over degradation of the liquefied starch. Glucoamylases can hydrolyze α-1,4 bonds rapidly but hydrolyze the α-1,6 bonds much more slowly. This obstacle is overcome with the combined use of glucoamylase and pullulanase. Pullulanase hydrolyzes the α-1,6 linkages rapidly, yielding straight-chain oligosaccharides that can be rapidly hydrolyzed by amyloglucosidase. The usage of these amylolytic enzymes in hydrolysis, either singly or in combination, has been found to be effective.

Since temperature, pH and other operation conditions are different depending upon the steps from gelatinization to saccharification, novel idea and optimization of the whole process steps are required when considering energy efficiency and economics.

13.1.4 Application to Textile Industry

Highly specific enzymes are becoming increasingly popular for various textile-processing applications because they can replace organic/inorganic agents. Pollution in the textiles industry is mainly related to the acids in cloths desizing process, bleaching agents, and dyes.

Removing starch paste from the fabric is called desizing, which used to be carried out by acid, alkali, or oxidizing agents, or by soaking in water for a long time. However, these methods can be hard to control and lead to damaging or discoloring the material. The problems could be overcome by introducing bacterial amylases, especially the thermostable ones. In contrary to chemical methods, enzymatic methods were able to lower the pollution while keeping the fabric quality. Another area where the traditional practice is replaced by enzymatic methods is fabric softening. Enzymatic methods give a softness maintained over a number of washes.

Furthermore, enzymes like cellulases, hemicellulases, lipases, proteases, and pectinases are used in the cotton scouring (Buchert *et al.* 1998). Sodium hydroxide, hydrogen peroxide, and sodium hypochlorite used to be employed in the process to remove the impurities from the raw cotton, but the enzymatic treatment offers an ecofriendly replacement; for example, cellulases have been reported to improve the cotton wettability. Additionally, the cellulose treatment before alkaline scouring could enhance the elimination of seed-coat fragments, improve whiteness of cotton fabrics, and reduce the consumption of hydrogen peroxide in the bleaching step (Csiszar *et al.* 1998). Cellulase, hemicellulase, pectinase, and xylanase can also help remove cotton seed-coat from the desized cotton. Other uses include degradation of hydrogen peroxide in the wastewater (catalase), dyeing step (redox enzyme systems), the prewashing step to increase dye absorption (cellulases, amylases, and proteases), or improving whiteness prior to dyeing, color shade and the felting (lignin peroxidase, manganese peroxidase and laccase).

Case Study: Isomaltooligosaccharides Production (Kim *et al.* 2002)

Maltose is being transformed into isomaltooligosaccharides (IMO) by specific α-glucosidases (E.C. 3.2.1.3; 1,4-α-d-glucan glucohydrolase). The exo-type enzymes hydrolyze amylose, amylopectin, and oligosaccharides as well as maltose. Interestingly maltose can be converted to isomaltose through coupling reaction via an α-1,6 linkage. Further glycosylation of isomaltose leads to isomaltotriose formation consequently.

Isomaltooligosaccharides are commercialized and sold as probiotic fibers in Japan. The α-glucosidase originated from *Aspergillus niger* is commonly used. 30% liquefied corn starch are converted to 68% IMO by adding α-amylase, pullulanase, and α-glucosidase at following typical conditions: 48 h, 58 °C, pH 5.5.

13.1.5 Brewing

Water, malted barley, hops, and yeast are necessary raw materials for brewing beer. The brewing process includes extraction and breaking the polysaccharides and protein mainly from malted barley. This process leads to aqueous solution containing plentiful sugars and proteins, which can be an ultimate resource for yeast fermentation.

Malted barley contains all the hydrolyzing enzymes required for the degradation of starch, β-glucans, pentosans, lipids, and proteins. If low quality malt or high fraction of adjunct is supplied as raw material, additional hydrolyzing enzyme will be beneficial. More freedom to choose diverse raw materials will be given to beer brewers by employing appropriate commercial enzymes aiming various beers.

Malting Process

For raw barley to be utilized for brewing beer, following pretreatment steps are required: (1) the seedling of barley grain to enhance the endogenous enzyme expression so called malting or (2) quenching of supplemented enzymes. The major purpose of malting is to stimulate the production of phytohormones and enzymes of the barley. For successful modified version of malting, several conditions must be met during malting:

- Sufficient breakup of endosperm cell wall is needed to supply more surfaces to be subjected to degradation by enzyme. Glucans must be degraded fully enough to restrain viscous wort.
- Sufficient amylose hydrolyzing enzymes must be generated to degrade starch of endosperm.

- Sufficient amino acids must be released during the hydrolytic degradation of proteins stored in malt to support the yeast growth during the fermentation.
- Undesired flavors originated from raw barley must be scattered and substituted with desired malt flavors.

Following three major steps including (1) steeping, (2) germination, and (3) kilning comprise malting process. Steeping process where barley is submerged in water at approximately 12–15 °C helps to elevate the moisture level of barley from 10 to ~45%. To avoid anaerobic condition air is frequently supplied to submerged grain. Maintaining aerobic state is necessary for active metabolism within the embryo. The main purpose of steeping procedure is to make embryo hydrated and alert its activities.

13.1.6 Enzymes for Other Food Applications

Enzymes for food applications can be extended to enzymes for fatty acids, vitamins, etc. Modification and hydrolysis of fatty acids are nowadays becoming more and more important because of health issues and to find useful components for cosmetics, food, and pharmaceuticals. For example, beta-carotene from plant can be converted to Vitamin A and Vitamin A analogs using enzymes. Many useful components can be found from enzymatic conversion of fatty acids.

13.2 Enzymes for Bioenergy

Bioethanol can be produced from materials containing carbohydrates. The raw materials for bioethanol are classified into three categories: sucrose containing feedstocks, starch, and lignocellulosic materials. Brazil utilizes sugarcane for bioethanol production while the United States mainly uses starch from corn.

13.2.1 Bioethanol from Starch

The production of bioethanol from starchy materials typically includes sequential processes such as gelatinization, liquefaction, and saccharification to produce glucose and then fermentation of glucose to yield bioethanol. During gelatin formation procedure, the excess water is mass-transported into the starch granule through diffusion causing swelling and resulting in a breakup of crystallinity (Evans and Haismann 1982). Liquefaction of starch is consequently followed by enzymatic hydrolysis using thermostable amylases (Aiyer 2005). Alpha-amylases (E.C.3.2.1.1) are endoamylases catalyzing the hydrolysis of internal α-1,4-glycosidic linkages in starch in a random manner (Nigam and Singh 1995). The second stage of enzymatic hydrolysis is saccharification. During saccharification, the products after liquefaction further breaks. Glucoamylases (E.C.3.2.1.3) or β-amylases (E.C.3.2.1.2) are the enzymes used for saccharification. The starch saccharifying enzymes catalyze the hydrolysis of α-1,4- and α-1,6-glycosidic bonds of starch from the nonreducing ends giving glucose as the final product. The yield of starch hydrolysis will be enhanced further if the saccharifying enzyme will be combined with pullulanase (E.C.3.2.1.41) or other debranching enzymes. Once the starch is converted to simple sugars, it is then fermented by microorganisms such as yeast to produce bioethanol. The most commonly used yeast strain is *Saccharomyces cerevisiae* due to its high ethanol tolerance. Fermentation of simple sugars is a process following the hydrolysis of starch in the overall ethanol production. Doing sequential steps requires additional equipment and time to complete the entire ethanol production process. It has also higher energy requirement since gelatinization and liquefaction are normally done at elevated temperature.

13.2.2 Bioethanol from Lignocellulose

Nowadays bioethanol from renewable and lignocellulosic resources such as wood, corn stover, rice straw, switch grass, miscanthus as well as spent furnitures and waste paper is primary interest. Lignocellulosic biomass is composed of three crucial constituents namely hemicelluloses, cellulose, and lignin. Pretreatment of this biomass to release the simple sugars is necessary for bioethanol production. Pretreatment process means any physical and chemical treatments to increase the solubilization of these constituents of biomass. This pretreatment is performed to deconstruct the biomass network to decline the degree of crystallinity of the cellulose, which is beneficial for enzymatic hydrolysis. The goals of pretreatment on lignocellulosic material are shown in Fig. 13.3. For an effective pretreatment process, there should be a direct or subsequent sugar formation during hydrolysis, less degradation and loss of produced sugars, less formation of inhibitory products, reduction of energy demands and cost is minimized.

Generally there are three types of enzymes that are required to hydrolyze cellulose materials into glucose monomers:

exo-1,4-β-glucanases (cellobiohydrolase) (e.g. EC 3.2.1.91 and EC 3.2.1.176),
endo-1,4-β-glucanases (e.g. EC 3.2.1.4) and
β-glucosidases (cellobiases) (e.g. EC 3.2.1.21).

Cellobiohydrolases hydrolyze the ends of cellulose chains while endo-glucosidases cleave the cellulose chains in the middle and the degree of polymerization is thus decreased.

Hemicellulose has a varied composition in comparison to cellulose thus it requires a large number of enzymes to hydrolyze it effectively to simple sugars. Table 13.2 presents some enzymes used to degrade lignocellulosic materials.

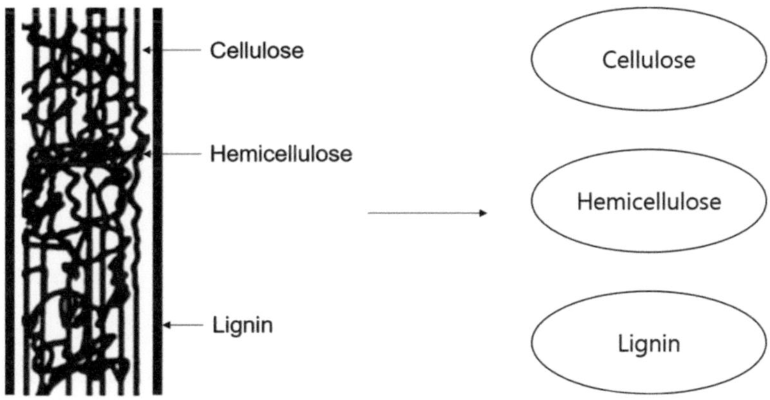

Fig. 13.3 Pretreatment of lignocellulosic material

Table 13.2 Enzymes used to degrade lignocellulosic materials

Component	Enzymes
Hemicellulose	Endo-xylanase, acetyl xylan esterase, β-xylosidase, endo-mannanase, β-mannosidase, α-L-arabinofuranosidase, α-glucuronidase, ferulic acid esterase, α-galactosidase, p-coumaric acid esterase
Cellulose	Cellobiohydrolyase, endoglucanase, β-glucosidase
Lignin	Laccase, manganese peroxidase, lignin peroxidase
Pectin	Pectin methyl esterase, pectate lyase, polygalacturonase, rhamnogalacturonan lyase

A lot of researches have been performed to reduce the cost of enzymes for bioethanol production from cellulosic materials, since enzyme cost is one of the dominating factor (Levine *et al.* 2011).

Biochemical conversion of lignocellulosic materials through saccharification and fermentation is a major pathway for bioethanol production. The use of lignocellulosic material shows economical and environmental advantages, however, bioconversion of lignocellulosic to bioethanol has intrinsic difficulties owing to: (1) the recalcitrant properties of biomass; (2) rare sugars such as xylose to require specific metabolism to be metabolized; (3) high costs for collection and transport of lignocellulosic materials.

Bioethanol production involves sequential steps such as pretreatment, hydrolysis of feedstock materials to release fermentable sugars, fermentation, separation, and concentration of the product ethanol. For these sequential steps, each requires a unit of equipment. For pretreatment, steam explosion, dilute acid percolation, or ammonia percolation methods can be used. Enzymatic hydrolysis of lignocellulose biomass to fermentable sugars and consequent fermentation to ethanol can also be done in one reaction vessel in the process called simultaneous saccharification and fermentation (SSF), as shown in Fig. 13.4. SSF has some inherent advantages,

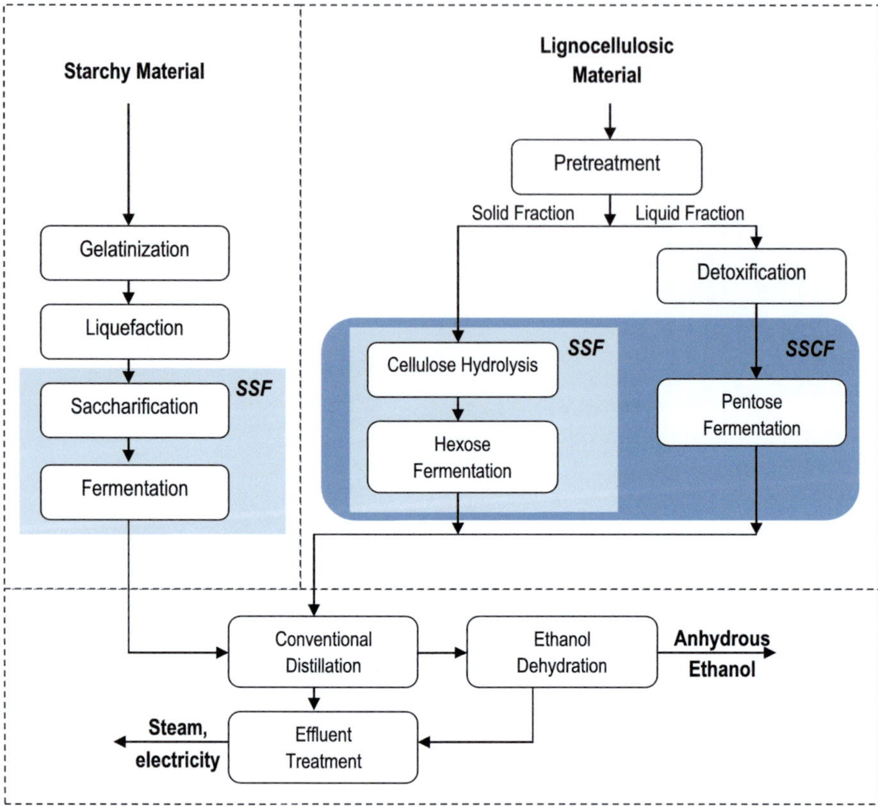

Fig. 13.4 Schematic diagram for bioethanol production

over separate hydrolysis and fermentation, such as the decrease in capital cost since an entire step could be eliminated. Glucose inhibition of the biocatalyst can be reduced since the sugars are directly metabolized after hydrolysis. One challenge of SSF is the different optimum temperature of the yeast used for fermentation and the enzyme used for saccharification. If microorganisms (e.g. genetically modified *Zymomonas mobilis*) capable of assimilating pentoses and hexoses are used, the simultaneous saccharification and cofermentation (SSCF) process could be applied. In SSCF, both fractions of the hemicelluloses hydrolyzate can be utilized for bioethanol production. Once the fermentation has been achieved, the culture broth is sent to the downstream processing to get the final product quality.

For economic production of bioethanol, several technologies are to be developed further, such as technology for cheap and efficient cellulase, utilization of lignin, glucose and xylose fermenting yeast, and ethanol separation with less energy. Another approach is to get bioenergy from algae. In this case, various polysaccharides are synthesized and thus should be hydrolyzed to monosaccharides which can be fermented by microorganisms. Enzymes are useful tools for the hydrolysis, such as laminarase in hydrolyzing laminarian to glucose.

13.2.3 Application to Pulp and Paper Industry

The pulp and paper industry uses harsh chemicals and bleaching compounds and produces deleterious by-products at many stages, causing major environmental pollution. Application of enzymes in the pulp and paper industry has a high potential in economics of the process and in alleviating environmental burden.

The paper manufacturing mainly involves three steps: pulping, pulp dewatering, and refining. Lignocellulosic raw material is digested with calcium sulfite and then dewatered and refined. Afterwards, the refined pulp is bleached with Cl_2 for whitening. Meanwhile, the residual lignin could be completely removed in the bleaching of pulp (Ahuja *et al.* 2004). The main focus areas of enzyme technology in the pulp and paper industry are as follows (Torres *et al.* 2012): (1) biopulping to save energy and replace harmful chemicals; (2) water treatment; and (3) to solve problems related to deinking, drainability, hornification, deposits of pitch and stickies, and biofilm formation, as shown in Fig. 13.5.

Generally, biopulping process makes use of the enzymes originated from fungi to reduce the consumption of chemicals in the pulping stage of wood chips, to

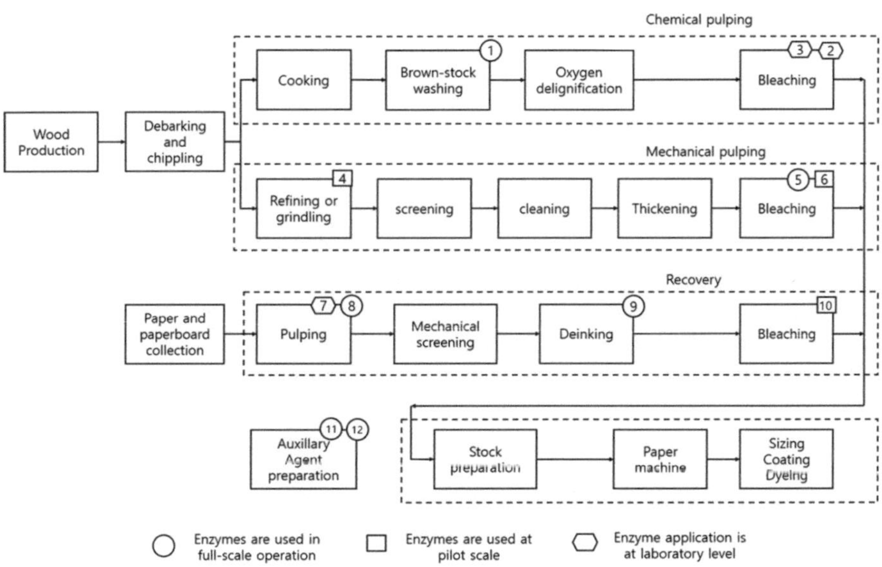

Fig. 13.5 Main processes in the pulp and paper industry with indication of present and potential points of enzyme application (Skals *et al.* 2008)

increase the yield of fiber, to reduce further refining energy, or to provide specific fiber modifications. The treatment of wood chips using enzymes could open up the cell wall and thereby separate the fiber at preferable locations in subsequent refining. The wood chips become softened and porous to be easily broken apart during pulping and specifically during refining. As wood chips are treated at high temperature and basic pH, the enzymatic procedures require enzymes exhibiting a high thermostability and activity in a broad pH range. Enzyme treatment of wood chips or coarse fibers in mechanical pulping may lead to significant energy saving, but it is still urgent to develop an applicable technology at industrial scale. The enzymes involved in biopulping include cellulases, xylanases, laccases, manganese peroxidases, amylases, and pectinases. If xylanases are used along with bleaching agents, the use of oxidizing chemicals can be reduced by 15–20%. If cellulases are used in deinking, usage of deinking chemicals can be reduced for environmental benefit.

13.2.4 Biodiesel

Origin of diesel was biodiesel from plant oil before cheap oil from petroleum was used. Corn oil, palm oil, or rapeseed oil can be used to make biodiesel. Fatty acid methyl ester (FAME) is a methyl esters of long chain fatty acids. Biodiesel is synthesized by esterification of fatty acids or by transesterification of triglycerides with short-chain alcohols as illustrated in Fig. 13.6. Vegetable oils, animal fat, waste cooking oil, greases, and algae can be used as feedstock materials for biodiesel production.

Biodiesel can be produced by chemical or biochemical catalysis method. Currently, the most common commercial process for biodiesel production is through chemical catalysis. Chemical catalysis offers high biodiesel yield, low catalyst cost, and high process productivity. However, soap formation, catalyst recovery after reaction, and wastewater treatment need are some of the problems associated for chemical method. Contrary to chemical catalysis, the enzymel method is now being considered due to the following merits: (1) it can be operated at milder reaction conditions, (2) more selection of feedstocks, (3) easier separation of the immobilized catalyst from the reaction mixture (4) easier for separation and purification of the product in downstream process, and (5) environmentally friendly processing. Recently, commercial production of biodiesel is being performed in China, where repeated use of immobilized lipase is known as one of the key technology.

Lipases can catalyze mono-, di-, and tri-glycerides as well as the free fatty acids (FFA) to yield FAME in organic solvent. A wide range of lipases has been used for esterification and transesterification processes since they exhibit low product inhibition, low reaction time, reuse of the enzyme, high stability to temperature and alcohol, and ease of lipase production.

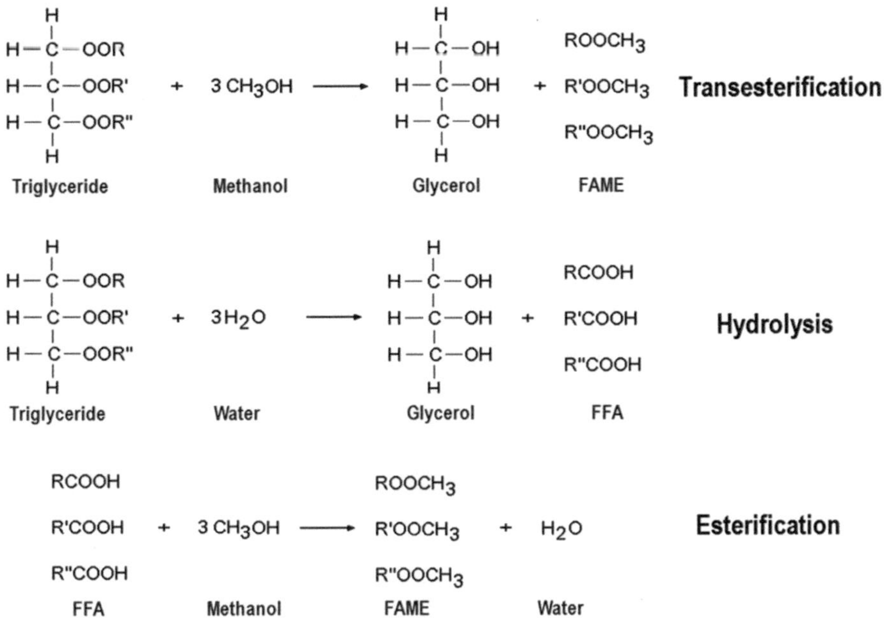

Fig. 13.6 Synthesis of biodiesel

13.2.5 Enzymes for Other Bioenergy

Even though bioethanol and biodiesel as popular bioenergy are currently produced and utilized, researches on other bioenergy production are required. Hydrogen can be produced by cyanobacteria, where nitrogenase and hydrogenase are involved. For efficient production of hydrogen from cyanobacteria, the enzymes involved are to be investigated in detail and engineered if required. Methane production from carbon dioxide and methanol from methane have also been important issues. Even though methane has been produced and utilized from anaerobic digestion, methane can be also obtained from different sources where the utilization as energy is remained as an important issue. To get ultimate bioenergy, deep understanding on photosynthesis is a challenging issue.

Further Discussion

1. What is gluten? What enzymes can be used to make gluten-free flour and gluten-reduced beer? Fatty acids are converted to functional ingredients by enzymes and beta-carotene can be also converted to Vitamin A and can be used as food supplement. Discuss on the use of enzymes in food industry.
2. Bioethanol is also being produced from sugar canes. Search the technology for producing bioethanol from sugarcanes.

3. Lignin is used as surfactant after chemical modification to sulfonates or used as energy fuel by burning the lignin. Lignin is to be treated or converted efficiently for economic production of bioethanol from cellulosic biomass. In nature, lignin is also degraded. Discuss the technology for efficient lignin treatment.
4. For biodiesel production using enzymes, what technology is important to produce the biodiesel for commercial purpose?
5. Bioethanol or biodiesel from algae is very important area. What technology are to be developed for bioenergy production from algae?
6. Can we produce methane from carbon dioxide, methanol from methane, glucose from carbon dioxide, water and light? What are the ultimate goals for bioenergy production?

References

Ahuja SK, Ferreira GM and Moreira AR. Utilization of enzymes for environmental applications. *Crit. Rev. Biotechnol.*, 2004, 24:125–154.

Aiyer, Amylases and their applications. *African Journal of Biotechnology*, 2005, 4(13):1525–1529.

Bekes F, Gras PW and Gupta RB. Mixing properties as measurement of reversible reduction and oxidation of dough. *Cereal Chemistry,* 1994, 71:44–50.

Boussard A, Cordella CBY, Rakotozafy L, Moulin G, Buche F, Potus J and Nicolas J. Use of chemometric tools to estimate the effects of the addition of yeast, glucose-oxidase, soybean or horse bean flours to wheat flour on biochemical, bread dough characteristics. *Chemometrics and Intelligent Laboratory Systems,* 2012, 113:68–77.

Buchert J, Pere J, Puolakka A and Nousiainen P. Enzymatic scouring of cotton. In: *Book of Papers, Proceedings of the AATCC International Conference and Exhibition*, American Association of Textile Chemistry and Color, Philadelphia, 1998, 493–499.

Collar C, Martilnoz J C, Andrew P and Armero E. Effect of enzyme association on bread dough performance: A response surface study, *Food Sci Technol Int*, 2000, 6:217–226.

Csiszar E, Szakacs G and Rusznak I. Combining traditional scouring with cellulase enzymatic treatment. *Textile Res. J.,* 1998, 68:163–167.

Evans and Haisman, The Effect of Solutes on the Gelatinization Temperature Range of Potato Starch. Starch - St rke, 1982, 34(7):224–231.

Foltmann BF. Chymosin: a short review on foetal and neonatal gastric proteases. *Scandinavian Journal of Clinical and Laboratory Investigation,* 1992, 52:65–79.

Harboe M and Budtz P. The production, action and application of rennet and coagulants. In: *Technology of Cheesemaking* (ed. B.A. Law). *Sheffield Academic Press*, 1999.

Levine SF, Fox JM, Clark DS and Blanch HW. A mechanistic model for rational design of optimal cellulose mixtures. *Biotechnology and Bioengineering*, 2011, 11:2561–2570.

Motoki and Seguro, Transglutaminase and its use for food processing. *Trends in Food Science and Technology*, 1998, 9(5):204–210.

Nielsen, P.H., Malmos, H., Damhus, T., Diderichsen, B., Nielsen, H.K., Simonsen, M., SchiV, H.E., Oestergaard, A., Olsen, H.S., Eigtved, P. and Nielsen, T.K. Enzyme applications (industrial). in*: fourth ed..Kirk-Othmer Encyclopedia of Chemical Technology,* John Wiley & Sons 1994, 567–620.

Nigam P and Singh D. Enzymes and microbial systems involved in starch processing. *Enzyme Microbiology and Technology,* 1995, 17:770–778.

Pariza and Johnson, Evaluating the safety of microbial enzyme preparations used in food processing: Update for a new century. *Regulatory Toxicology and Pharmacology*, 2001, 33(2):173–186.

Rouau X. Investigations into the effects of an enzyme preparation for baking on wheat flour dough pentosans, *J. Cereal Sci.* 1993, 18:145–157.

Skals PB, Krabek A, Nielsen PH and Wenzel H. Environmental assessment of enzyme assisted processing in pulp and paper industry. *Int. J. LCA*, 2008, 13:124–132.

Torres CE, Negro C, Fuente E and Blanco A. Enzymatic approaches in paper industry for pulp refining and biofilm control. *Appl. Microbiol. Biotechnol.*, 2012, 96:327–344.

T.W., Kim, J.W. and Park, K.M. Cooperative action of alpha-glucanotransferase and maltogenic amylase for an improved process of isomaltooligosaccharide (IMO) production. *Journal of Agricultural and Food Chemistry,* 2002, 50:2812–1817.

Van Der Maarel, M.J.E.C., Van der Veen, B.A., Uitdehaag, J.C.M., Leemhuis, H. and Dijkhuizen, L. Properties and applications of starch-converting enzymes of the alpha-amylase family. *Journal of Biotechnology,* 2002, 94:137–155.

Chapter 14
Enzymes for Biosensors

14.1 Enzymatic Measurements

A biosensor is defined by the National Research Council of the USA as a detection device that incorporates (1) a living organism or product derived from living systems (e.g., an enzyme or an antibody) and (2) a transducer to provide an indication, signal, or other form of recognition of the presence of a specific substance in the environment. Biosensors must be able to detect molecules of analytical significance fast, accurately and reliably. Biosensors have important roles in environmental and industrial monitoring, medicine, agriculture, food, security, and bioprocessing. Significant improvements in biosensors' selectivity and detection sensitivity have been made to facilitate their applications. Many different biosensor-based transducers (see Table 14.1) have been developed and those are grouped as electrochemical (impedance-based, amperometric), optical (fiber optic, surface plasmon resonance), thermometric (thermistor, pyroelectric), and mass based (piezoelectric, surface acoustic).

Enzyme sensors are a major part of biosensor technology, which currently represent a mature analog to instrumental analytical techniques in areas of health care, food industry, agricultural issues, as well as environmental monitoring. Enzymes possess high chemical specificity and have natural bio-catalytic signal amplification property. In an enzymatic sensing device, enzyme is combined with a transducer where it reacts selectively with its analyte and produces a signal corresponding to the target analyte concentration, in the form of proton concentration, release or uptake of gases, light emission, absorption or reflectance, heat emission, and so forth. The conversion of this signal into a measurable response is performed by a transducer. Since enzyme-based bio-analytical devices offer several distinct advantages such as high sensitivity and specificity, portability, cost-effectiveness, and the possibilities for miniaturization and mass production, it has been

Y.J. Yoo *et al.*, *Fundamentals of Enzyme Engineering*,
DOI 10.1007/978-94-024-1026-6_14

Table 14.1 Examples of transducers

Bio-recognition elements	Transducers	Measured property
Enzymes, proteins, amino acids nucleic acids: DNA, RNA, PNA, antibodies, antigens, specific genes, organelles, microbial cells, plant and animal tissues	Electrochemical	Potentiometric, conductometric, field effect amperometric voltammetric impedimetric Surface conductivity, electrolyte conductivity
	Optical	UV absorption, fluorescence emission, optical quantitative imaging adsorption, bioluminescence, chemiluminescence, optical grating coupler sensing technology, reflection, internal reflection spectroscopy, laser light scattering, grating based light diffraction, imaging ellipsometry
	Mass sensitive	Resonance frequency of piezocrystals, piezoelectric acoustic wave modes, quartz crystal, cantilever sensor technology
	Thermal	Heat of reaction, heat of adsorption, thermistor sensor

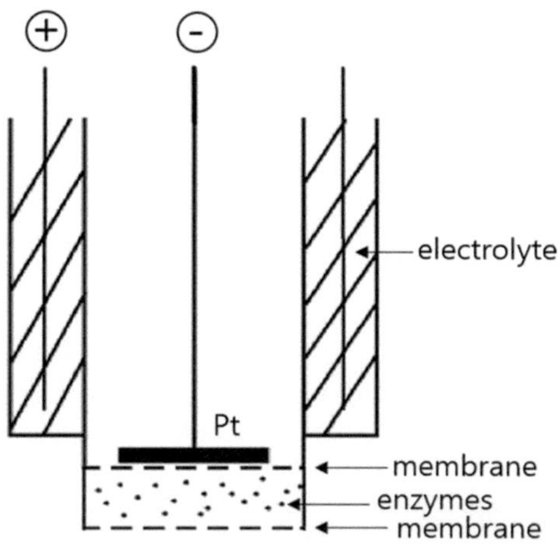

Fig. 14.1 Schematic diagram of glucose biosensor employing DO probe

extensively implemented in different aspects, some of which are briefly introduced in this chapter.

Glucose Sensor. Glucose sensor is widely used in hospital and industry and has been extensively developed from long time ago. When substrate or analyte reacts with enzyme-yielding product, the product can be measured using many methods, and then substrate concentration can be calculated. Figure 14.1 shows the scheme for typical and classical glucose sensor, where dissolved oxygen probe (galvanic or potentiometric type) is a part of glucose sensor.

Enzyme reactions

$$\text{glucose} \underset{\text{glucose oxidase}}{\longrightarrow} \text{gluconic acid} + H_2O_2$$

$$H_2O_2 \underset{\text{catalase}}{\longrightarrow} H_2O + O$$

$$\text{Cathode: } 1/2O_2 + H_2O + 2e^- \underset{\text{Pt}}{\longrightarrow} 2OH^-$$

$$\text{Anode (galvanic type): } Pb \longrightarrow Pb^+ + 2e^-$$

$$\text{Anode (polarographic type): } Ag + Cl^- \longrightarrow AgCl + e^-$$

When combining dissolved oxygen probe with immobilized enzymes, the electrons released are in proportional to the glucose level within the calibration range.

Sensing Methods and Transducers

Electrochemical Sensors. In electrochemical sensors, pH changes, ions are formed or oxygen is consumed as a result of enzyme reaction with the substrate, which generate electrical signals on a transducer. Potentiometric, capacitive, and amperometric transducers have been used for such applications. One example is the glucose sensor. Alkaline phosphatase (AP) hydrolyzes *p*-nitrophenyl phosphate to phenol, which is detected by voltammetry. In light-addressable potentiometric sensors (LAPS), urease hydrolyzes urea to produce carbon dioxide and ammonia that change the pH.

Optical Biosensor. Fiber optic biosensor mechanism relies on fluorescently labeled analytes which are excited by laser upon binding to the surface of the biosensor. The excitation leads to fluorescent signals detected in real time. Optical biosensors have been utilized to detect various pathogens and contaminants in food. In many cases, fluorophore-labeled antibodies are used to specifically bind to the analytes. FITC (fluorescein isothiocyanate) is the most widely used, followed by some lanthanides (Li *et al.* 2004). The labels are also used in PCR or ELISA. Fiber optic biosensor is one of the first commercially available optical biosensors which was used to detect food-borne and pathogens of biosecurity importance.

Surface Plasmon Resonance Biosensor. In surface plasmon resonance (SPR) sensor, antibodies to capture various pathogenic analytes are coated on thin gold film covering reflecting surface of waveguide. The sensing surface is located above or below a high index resonant layer and a low index coupling layer. When a visible light or a near-infrared (IR) light pass through the waveguide, an internal total reflection on the surface occurs. High resonance due to an interaction of the light with the electron cloud in the metal appears at specific wavelength. When an analyte binds to the metal surface via the antibodies, a shift in the resonance to longer wavelengths is observed. The amount of shift is correlated with the amount of the binding analyte. SPR biosensors have detection limits of femtomolar range (Banada *et al.* 2007; Rasooly and Herold 2006).

Piezoelectric Biosensor. Piezoelectric sensors are mass sensitive. When an analyte binds to a specific antibody immobilized on the surface of the sensors, increased mass of a quartz crystal causes a proportional decrease in the oscillation frequency of the crystal, to be detected by the quartz crystal microbalance (QCM) on a transducer (O'sullivan and Guilbault 1999). Lithium niobate is a promising alternative to the quartz (Leonard *et al.* 2003). Piezoelectric biosensors are simple yet very useful real-time technique for detection of food-borne pathogen.

14.2 Applications of Enzyme Biosensor

For Biomedical Analysis. Enzymatic biosensors broadly used in the biomedical sector are used to detect clinical biomarkers such as cholesterol, glucose, glutamate, lactate, and urea. Disposable blood glucose sensor is by far the most extensively studied and commercialized due to the large population of patients with diabetes and other metabolic disorders. Enzymatic biosensors for the real-time detection of brain dopamine have also been reported as a replacement to the conventional fast scan cyclic voltammetry. Polyphenol oxidase, with an activity on dopamine, is utilized to produce dopaquinone detectable electrochemically at -150 mV (Njagi *et al.* 2008). Recently, microfluidic paper-based platforms have been combined with traditional electrochemical readers. The integrated device allowed rapid quantitative detection of clinical analytes; for example, glucose, cholesterol, lactate, and alcohol in blood and urine (Nie *et al.* 2010). Moreover, it has been reported the use of paper-based screen-printing electrodes for detection of glucose, lactate, and uric acid in human serum samples using their respective oxidase enzymes (Dungchai *et al.* 2009).

For Agricultural and Food Industry. Biosensors are used to check the freshness of raw foods and to control quality and safety during the production. In these cases, sensing appliances are used to measure various compositions and evaluate rancidity, maturity, decline, and shelf life of the food.

For instances, quantification of ethanol and methanol in foods is carried out to assess the food liveliness and the quality of alcoholic beverages. Enzymatic sensing tools have been constructed, mostly using the alcohol dehydrogenase and oxidase, and catalase less commonly (Valach *et al.* 2009). Alcohol oxidase biosensor to detect ethanol is the most abundant. Smyth *et al.* (1999) reported a biosensor in which an alcohol oxidase was co-immobilized with an alcohol peroxidase and a chromogen, for detection of damages in vegetables processed in low O_2 environment. In addition, this biosensor could be applied where ethanol accumulation is related with a quality loss, such as the storage of apples in a controlled atmosphere and the decay in potato tubers (Castillo *et al.* 2003). Detection of organic acids and sugars as a measure to fruit and vegetable maturity and other food contaminants has also been investigated (Cnas and Macias 2004). Noguer *et al.* (2001) constructed a disposable aldehyde dehydrogenase (AlDH) sensor to detect MITC (methyl isothiocyanate), a metabolite of metam sodium. Benzoic acid in mayonnaise and soft drinks could be determined via a graphite–Teflon–tyrosinase composite biosensor, with a low detection limit up to 9×10^{-7} mol^{-1}, good renewability, high stability, and low cost (Morales *et al.* 2002).

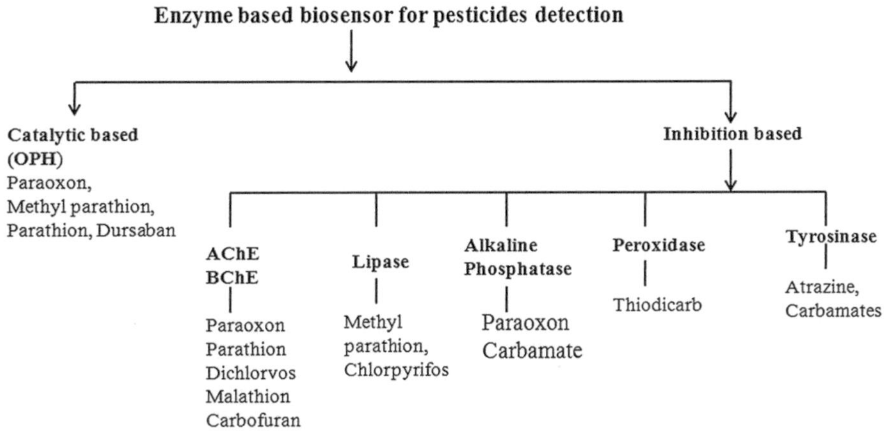

Fig. 14.2 Classification enzyme-based biosensors for the pesticide analysis. *AChE* acetylcholinesterase, *BChE* butyrylcholinesterase, *OPH* organophosphorus hydrolase (Neelam and Atul 2015)

For Pesticide Detection. Herbicides, insecticides, fungicides, and rodenticides are heavily utilized in agriculture and need to be monitored by the industry, health care professionals, and regulatory agencies. Two methods dictate enzymatic biosensors for pesticides: direct technique where the enzymatic reaction leads to a change in the amount of measurable chemical, and indirect technique which monitors inhibition on the enzyme. Classification of enzyme based on pesticide biosensors is presented (Fig. 14.2).

Inhibition Biosensor. Many analytes such as organophosphorous pesticide (OP), organochlorine pesticides, carbamate pesticides, and derivatives of insecticides are inhibitors of various enzymes. Several enzymes such as acetylcholinesterase (AChE), butyrylcholinesterase (BChE), and urease are applied to make direct electrochemical biosensors for the inhibitory pesticides. In addition to a single enzymatic sensor, multi-enzyme sensing devices have been developed such as an integrated sensor where AChE and BChE are co-immobilized on the carbon-based ink-printed electrodes. Chlorfenvinphos and diazinon, organophosphate pesticides used in wool, could be determined by this type of biosensor (Collier *et al.* 2002).

Catalytic Biosensors. Another approach for detection of pesticides is to make use of the organophosphorus hydrolase (OPH), a type of bacterial phosphotriesterase, with a broad substrate specificity for various OP derivatives such as paraoxon, parathion, coumaphos, diazinon, dursban, and methyl parathion, and nerve gas like sarin and soman (Dumas *et al.* 1990). A number of OPH-based amperometric, potentiometric, and optical devices have been constructed (Mulchandani *et al.* 1999; Viveros *et al.* 2006; Choi *et al.* 2010). Using genetically engineered microbes as sources of OPH enzyme, whole cells have also been utilized as biosensors for the OP pesticides (Mulchandani *et al.* 2001).

For Monitoring Other Environmental Pollutants. A wide range of enzyme-based sensing devices have been developed for environmental analysis. To date, heavy metal contamination is a very common environmental issue. However, to determine the heavy metal in various water sources, a large number of ion sensing devices with low detection limits have been constructed. For instances, horseradish peroxidase, glucose oxidase, invertase- and urease-based sensors for mercury, glutathione-*S* transferase for captan (Choi *et al.* 2003), and carboxyl esterase for selenium (Saritha and Kumar 2001) have been designed. Urease-based sensors are suitable for copper and cadmium in addition to mercury in tap and river water (Tsai and Doong 2005). Furthermore, different oxidases are used for nitric oxide (NO) (Kilinc *et al.* 2000) and superoxide radicals (Campanella *et al.* 2000) detection.

More Issues for Enzyme Biosensor

There are many issues in developing enzyme biosensor such as What's the sensing range? How about the sensitivity? How long can we use the sensor? How small? How to calibrate? How about the cost? Many sensors from chemical principles, physical principles, and microbial principles are competing with enzyme biosensor. However, due to advantageous figure of enzyme-based biosensing appliances, researchers are paying more attention to generate more advanced devices such as nano-sensors, paper-based sensors, lab-on-a-chip, biochips, and microfluidic devices to meet growing demands of the world. The technology on enzyme biosensor can also be applied to enzyme-based biofuel cells, which has several positive advantages (Minteer *et al.* 2007).

Further Discussion

1. Glucose monitoring from humans is very important. However, noninvasive glucose monitoring is desirable instead of currently used invasive method. What technologies are being developed for this purpose?
2. Biosensor for organophosphate pesticide or nerve gas detection is very important. Discuss the measurement range and sensitivity of the reported enzyme sensor for organophosphate detection.
3. When using enzymes for biosensing, enzyme deactivation during storage is important and should be considered for accurate measurement. What methods are being considered to overcome this instability of enzymes in biosensing?
4. Developing enzyme fuel cells is also an important issue. What is different in principle from enzyme biosensor? What idea from enzyme sensor technology can be used to develop enzyme fuel cell?

References

Banada PP, Guo S, Bayraktar B, Bae E, Rajwa B, Robinson JP, Hirleman ED and Bhunia AK. Optical forward-scattering for detection of *Listeria monocytogenes* and other *Listeria* species. *Biosensors and Bioelectronics*, 2007, 22(8):1664–1671.

Campanella L, Persi L and Tomasseti M. A new tool for superoxide and nitric oxide radicals determination using suitable enzymatic sensors. *Sens. Actuators B.*, 2000, 68:351–359.

Castillo J, Gáspár S, Sakharov I and Csöegi E. Bioenzyme biosensors for glucose, ethanol and putrescine built on oxidase and sweet potato peroxidase. *Biosensors and Bioelectronics*, 2003, 18(56):705–714.

Choi JW, Kim YK, Song SY, Lee IH and Lee WH. Optical biosensor consisting of glutathione-*S*-transferase for detection of captan. *Biosensors and Bioelectronics*, 2003. 18:1461–1466.

Choi BG, Park H, Park TJ, Yang MH, Kim JS, Jang SY, Heo NS, Lee SY, Kong J, and Hong WH. Solution chemistry of self-assembled graphene nanohybrids for high-performance flexible biosensors. *ACS Nano*, 2010, 4:2910–2918.

Cnas A, and Macias M. Desarrollo deunsistema sensor para la cuantificación deglucosa enjugos defrutas. *Revista de la Sociedad Química de Mexico*, 2004, 48(8):106–110.

Collier WA, Clear MA and Hart AL. Convenient and rapid detection of pesticides of extracts of sheep wool. *Biosensors and Bioelectronics*, 2002, 17:815–819.

Dungchai W, Chailapakul O, and Henry CS. Electrochemical detection for paper-based microfluidics. *Anal. Chem.*, 2009, 81:5821–5826.

Dumas DP, Durst HD, Landis WG, Raushel FM, Wild JR. Inactivation of organophosphorus nerve agents by the phosphotriesterase from Pseudomonas diminuta. *Archives of Biochemistry and Biophysics*, 1990, 277(1):155–159.

Kilinc E, Ozsoz M and Sadik OA. Electrochemical detection of NO by inhibition on oxidase activity. *Electroanalysis*, 2000, 12:1467–1471.

Leonard P, Hearty S, Brennan J, Dunne L, Quinn J, Chakraborty T and O'Kennedy R. Advances in biosensors for detection of pathogens in food and water. *Enz. Microb. Technol.*, 2003, 32:3–13.

Li Y, Dick WA and Tuovinen OH. Fluorescence microscopy for visualization of *soil* microorganisms-a review. *Biol. Fert. Soils*. 2004, 39:301–311.

Minteer SD, Liaw BY and Cooney MJ. Enzyme-based biofuel cells. *Current Opinion in Biotechnology*, 2007, 18:228–234.

Morales MD, Morante S, Escarpa A, Gonzalez MC, Reviejo AJ and Pingarron JM. Design of a composite amperometric enzyme electrode for the control of the benzoic acid content in food. *Talanta*, 2002, 57:1189–1198.

Mulchandani A, Mulchandani P, Chen W, Wang J and Chen L. Amperometric thick-film strip electrodes for monitoring organophosphate nerve agents based on immobilized organophosphorus hydrolase. *Analytical Chemistry*, 1999, 71(11):2246–2249.

Mulchandani P, Chen W, Mulchandani A, Wang J and Chen L. Amperometric microbial biosensor for direct determination of organophosphate pesticides using recombinant microorganism with surface expressed organophosphorus hydrolase. *Biosensors and Bioelectronics*, 2001, 16:433–437.

Neelam V and Atul B. Biosensor technology for pesticides- A review, *Appl Biochem Biotechnol.*, 2015, 175:3093–3119.

Nie Z, Deiss HF, Liu XY, Akbulut O, and Whitesides GM. Integration of paper-based microfluidic devices with commercial electrochemical readers. *Lab Chip*, 2010, 10(22):3163–3169.

Njagi J, Ispas C, and Andreescu S. Mixed ceria-based metal oxides biosensor for operation in oxygen restrictive environments. *Anal. Chem.*, 2008, 80(19):7266–7274.

Noguer T, Balasoiu AM, Avramescu A and Marty JL. Development of disposable biosensor for the detection of metam-sodium and its metabolite MITC. *Anal. Lett.*, 2001, 34(4):513–528.

O'sullivan CK and Guilbault GG. Commercial quartz crystal microbalances – theory and applications. *Biosens. Bioelectron.*, 1999, 14(8–9):663–670.

Rasooly A and Herold KE. Biosensors for the analysis of food-borne pathogens and their toxins. *J AOAC Int*, 2006, 89:873–883.

Saritha K and Kumar NVN. Qualitative detection of selenium in fortified soil and water samples by a paper chromatographic—carboxyl esterase enzyme inhibition technique. *J. Chromatogr.*, 2001, 919:223–228.

Smyth A, Talasila P and Cameron A. An ethanol biosensor can detect low-oxygen injury in modified atmosphere packages of fresh-cut produce. *Postharvest Biology and Technology*, 1999, 5:127–134.

Tsai HC and Doong RA. (2005) Simultaneous determination of pH, urea, acetylcholine and heavy metals using array-based enzymatic optical biosensor. *Biosensors and Bioelectronics*, 20:1796–1804.

Valach M, Katrlik J, Sturdik E, and Gemeiner P. Ethanol *Gluconobacter* biosensor designed for flow injection analysis: Application in ethanol fermentation off-line monitoring. *Sensors and Actuators B: Chemical*, 2009, 138:581–586.

Viveros L, Paliwal S, McCrae D, Wild J and Simonian A. A fluorescence-based biosensor for the detection of organophosphate pesticides and chemical warfare agents. *Sensor and Actuators B: Chemical*, 2006, 115:150–157.

Chapter 15
Enzymes for Environment

15.1 Enzymatic Removal of Pollutants

Strict regulations on the disposal of wastes to the environment require improvement of waste treatment processes. In recent years, extensive research on biological processes has been conducted to enable industrial, agricultural, municipal, and commercial facilities to reduce their harmful impacts on the environment. Biological processes such as activated sludge process are the most economical method when treating broad range of compounds in aqueous solution. They can be conducted under mild conditions. Biological processes use the innate cell metabolism to convert chemical species into harmless and clean metabolites. These metabolic processes occur as a result of sequential enzymatic reactions inside the cell. However, metabolism of the microorganisms limits the rate of the reaction. Moreover, microorganisms require a supply of nutrients to support growth and often lead to excessive biomass quantity that must be treated again for the environment. Since biological processes are not selective in reducing toxic pollutants and not efficient enough in treating high concentration of pollutants, they may not be sufficient to improve water quality that meets increasingly strict discharge criteria.

Selective removal of a specific chemical compound may be very important to facilitate subsequent treatment (e.g., biological processes, since microorganisms can be inhibited by toxic substances) or to meet regulatory criteria upon disposal. Isolated enzymes have high selectivity which can be efficiently used to target-specific pollutants for treatment. Enzymatic processes have higher reaction rates compared to biological process and lower quantity of sludge production since biomass is not generated. Enzymatic treatment might be beneficial for the following applications:

Y.J. Yoo *et al.*, *Fundamentals of Enzyme Engineering*,
DOI 10.1007/978-94-024-1026-6_15

(1) removal of a specific toxic compound from a complex waste mixture,
(2) removal of high concentrations of chemicals from waste stream for which conventional mixed culture treatment might not be possible,
(3) treatment of wastes generated infrequently or in isolated locations.

There are many enzymes that have been known for the treatment of waste materials that includes aromatic compounds, cyanide, color-causing compounds, pesticides, surfactants, and heavy metals. Some of these applications are described below.

Enzymes can be directly used in pollutant removal to improve the efficiency in waste treatment plants. To date, there have been many reports on the use of enzymes in soil bioremediation, in the detoxification of industrial wastewater and the removal of phenol from coal conversion aqueous effluents using horseradish peroxidase, and enzymes in detoxifying pesticide-contaminated soils and waters.

15.1.1 Aromatic Pollutants

Phenolic and aromatic amine compounds are pollutants regulated in many countries. They are normally contained in the wastewater of various industries such as coal conversion, petroleum refining, wood preservations, metal casting, resins and plastics, textiles, and pulp and paper manufacturing. Most of these compounds are classified as hazardous and toxic pollutants, and thus it is necessary that those compounds should be removed from wastes before discharging into the environment.

Horseradish peroxidase (HRP) can oxidize phenols, biphenols, anilines, benzidines, and other heteroaromatic compounds. HRP is suitable for wastewater treatment because it has activity over a broad pH and temperature range. HRP treatment was proposed by Klibanov, which has been used to remove over 30 different phenols and aromatic amines from water with removal efficiencies for some pollutants exceeding 99% (Klivanov and Morris 1981). HRP catalyze the oxidation of aromatic compounds with hydrogen peroxide which yields phenoxyl radicals. These radicals couple to form insoluble oligomers and can be separated by filtration. However, the enzyme has short catalytic lifetime attributed to the inactivation of the peroxidase during reaction.

Enzyme immobilization is one of the methodologies that improve useful life of the enzymes. Tatsumi *et al.* (1996) studied the clearance of diverse chlorophenols (10–200 μM) from wastewater using immobilized HRP, reducing 100% of total organic carbon (TOC) and 90% of adsorbable organic halogen (AOX). The immobilized enzyme was more efficient than the free one, and physical adsorption on magnetite was better than cross-linking in terms of the immobilization method. Immobilized HRP on activated silica resulted in 37% decolorization, 60% mineralization, and no loss of activity after being frozen for 2 months or after 5 days in kraft effluent (Dezott *et al.* 1995).

Cyanide Wastes

Cyanides are used in different industrial processes including the production of chemical intermediates, rubber, synthetic fibers, and pharmaceuticals. Since cyanide is a metabolic inhibitor toxic to humans and other organisms, its removal from effluents prior to discharge is essential.

Cyanidase enzyme converts cyanide into ammonia and formate in a probable single-step reaction. Basheer *et al.* (1993) studied the detoxification of a cyanide-containing extract from debittering apricot seeds by cyanidase in the food industry. In this report, immobilized cyanidase was packed in a diffusional-type, flat-membrane reactor for protecting enzymes from adverse effects of high molecular components contained in the extract, and thus the cyanidase retained its original activity for more than 400 h on steam. Different from cyanidase, cyanide hydratase has been reported to hydrolyze cyanide to formamide and remove the toxic HCN (Dumestre *et al.* 1997). Such enzymes can be immobilized and used in packed-cell columns to continuously degrade cyanides.

Pesticide Residues

Pesticides are widely used for crop protection and these include herbicides, insecticides, and fungicides. Common treatment methods for pesticides residues include incineration, chemical methods, and landfilling. However, these methods are of high cost and hazardous byproducts are also produced. Besides that, disposal of chemical reagents and the susceptibility of sensitive biological treatment system are other problems to be addressed.

Synthetic organophosphate compounds are highly toxic due to their irreversible inhibition of central nervous system acetylcholinesterase in all vertebrates. The continuous and excessive use of these compounds as agricultural pesticides has led to the contamination of many terrestrial and aquatic ecosystems. One of the main components of warfare agent, nerve gas, is also organophosphate, and detection and detoxification of the compounds are also very important for security purpose. Enzymatic degradation of organophosphate compounds has thus received considerable attention because both environmentally friendly and *in situ* detoxification are possible. Parathion hydrolase has been proposed for the detoxification of organophosphate pesticides, being able to hydrolyze methyl and ethyl parathion, diazinon, fensulfothion, dursban, and coumaphos (Caldwell and Raushel 1991). Recently, Feng's group has explored a new type of phosphotriesterase (PTE) from thermophile *Geobacillus kaustophilus* HTA426 which could proficiently hydrolyze various lactones and possessed a weak PTE activity, and the enzymatic activity against organophosphate pesticides could be dramatically improved through crystallographic analysis and protein engineering techniques (Dumas *et al.* 1990; Zheng *et al.* 2011; Zhang *et al.* 2012).

Case Study: Removal of Phenol by HRP

Phenolic compounds and direct dyes are widely distributed in the wastewater of the textile industry and they can be highly harmful toward aquatic life and humans. Up to now, a number of techniques including adsorption, chemical oxidation, solvent extraction, and biodegradation have been developed to remove the phenolic compounds and dyes from wastewaters. Among these treatment techniques, the use of oxido-reductive enzymes such as horseradish peroxidase (HRP) to catalyze the removal of pollutes has become increasingly important.

As HRP could catalyze the oxidation of phenols, biphenols, anilines, benzidines, and related heteroaromatic compounds, it is suitable for wastewater treatment over a broad pH and temperature range. To further improve the enzyme operational stability and reduce the treatment cost, enzyme immobilization has been widely employed through different supports. For example, immobilized HRP on modified acrylonitrile copolymer membrane could achieve a high degree of phenol oxidation (95.4%) in phenol solution with 100 mg/L concentration, and only 50% deactivation of immobilized enzymes was observed after the 20th day of the enzyme operation (Vasileva *et al.* 2009). Other supports such as sodium alginate and phospholipid-templated titania have also been used for constructing immobilized HRP, and these enzymes exhibited good removal efficiency of multiple phenolic compounds and operational stability (Alemzadeh and Nejati 2009; Jiang *et al.* 2014).

Since peroxidase is rapidly inactivated caused by phenoxyl radicals formed during oxidative polymerization, engineering of peroxidase can be a solution for stable operation. Engineering of the peroxidase was successfully performed to retard the rapid inactivation of the wild-type peroxidase and can be thus efficiently used for the removal of phenolic compounds from industrial wastewater (Kim *et al.* 2014).

Enzymes in Leather Industry

In leather manufacturing, a large amount of waste material is generated from animal hides and skins. Generally, six steps are involved: curing, soaking, deharing, dewooling, bating, and tanning. The discharges and refuges disposed from all these steps in the leather production will cause severe health hazards and environmental problems to the entire eco-system. The industrial effluents mainly contain higher amount of sulfide and chromium for improving the quality of tanning. The utilization of enzymes in the process has proved to be useful for both enhancing the leather quality and decreasing the pollutants (Choudhary *et al.* 2004).

Dehairing and dewooling are the largest process in leather production, which requires the involvement of industrial enzymes like proteases, amylases, and lipases (Thanikaivelan *et al.* 2004). In the traditional procedure, these two steps need an extremely alkaline environment in the soaked and swollen epidermis and corium of the skin, and then breaking the bonds of hair protein fibrils and dissolving the proteins of hair root by the reaction with sulfides. Employing protease in these steps could reduce the use of sulfides by 40%. Compared with lime and sodium sulfide that can degrade the hair and open up the fiber structure, enzymes do not dissolve the fiber and thus the hair can be removed easily from the liming float via filtration. Therefore, the chemical and biological oxygen demands could be dramatically reduced.

Bating is a key step to enhance the quality of leathers; stiff leather from light bating is used for soles, compared to soft leather from intense bating which is used for gloves. Compared to traditional process, introduction of trypsin is reliable and can lead to a lower biological oxygen demand effluent. Degreasing of the leather before tanning, which is the removal of fat, is also very important for the effects of following chemical processes such as tanning, retanning, and dyeing if most of the natural fat has been removed. Lipases are environment-friendly means of removing the fat to ensure the final quality of product, and meanwhile it can allow the partial or full replacement of the harmful solvents or surfactants.

15.1.2 Nitrogen Removal

Removal of nitrogen from wastewater is one of the processes that utilizes biological means. Nitrogen compounds, such as ammonia and nitrate, are toxic to aquatic species and causes eutrophication in aquatic environments. They can be effectively removed through biological nitrification-denitrification. A number of nitrogen removal processes (e.g., simultaneous nitrification and denitrification, partial nitrification and denitrification, and anaerobic ammonium oxidation) have been developed to facilitate nitrification and denitrification processes. When looking into the mechanism of nitrification and denitrification, we can find that the mechanisms are based on enzyme reactions. For more efficient treatment, it is desirable to understand the enzyme reactions inside of microorganisms.

Conventional microbial nitrogen removal is based on autotrophic nitrification and heterotrophic denitrification (Jetten *et al.* 2001). Nitrification process involves two steps: (1) Membrane-bound ammonia monooxygenase (AMO) and hydroxylamine oxidoreductase (HAO) in ammonia-oxidizing bacteria (AOB) oxidize ammonium $\left(NH_4^+\right)$ to nitrite $\left(NO_2^-\right)$ via hydroxylamine (NH_2OH) and (2) membrane-bound nitrite oxidoreductase (NOR) in nitrite-oxidizing bacteria (NOB) oxidizes nitrite to nitrate $\left(NO_3^-\right)$:

$$\overbrace{NH_4^+ \xrightarrow[E_1]{} NH_2OH \xrightarrow[E_2]{} NO_2^-}^{AOB} \overbrace{\xrightarrow[E_3]{} NO_3^-}^{NOB} \tag{15.1}$$

where E_1: AMO, E_2: HAO, E_3: NOR

In denitrification, nitrate and nitrite are reduced to gaseous nitrogen with four steps of enzyme reactions. When using microorganisms, a variety of electron donors such as methanol or acetate are to be supplied which gives burden as a cost:

$$NO_3^- \xrightarrow[E_1]{} NO_2^- \xrightarrow[E_2]{} NO \xrightarrow[E_3]{} NO_2^- \xrightarrow[E4]{} N_2, \tag{15.2}$$

where E_1: nitrate reductase, E_2: nitrite reductase, E_3: nitric oxide reductase, and E_4: nitrous oxide reductase.

Partial nitrification–denitrification process. In this process, nitrification and denitrification are correlated by NO_2^- instead of NO_3^-. Compared with the traditional nitrification–denitrification process, it has the following advantages:

$$NH_4^+ \longrightarrow NO_2^- \longrightarrow N_2 \tag{15.3}$$

(1) Lower oxygen consumption in the aerobic phase which implies lower energy requirement in the entire process;
(2) Lower requirement for electron donors in the anoxic phase; and
(3) Higher NO_2^- denitrification rate than NO_3^- denitrification rate.

Anaerobic ammonium oxidation* (ANAMMOX) *process. In ANAMMOX, anaerobic AOB oxidizes ammonia to nitrogen with nitrite as the electron acceptor. External carbon sources are not needed for the anaerobic AOB because carbon dioxide serves as the main carbon source (Jetten *et al.* 2001; Hu *et al.* 2013). The biomass yield for this process is very low which saves sludge treatment costs (Hu *et al.* 2011)

$$NH_4^+ \longrightarrow NO_2^- \tag{15.4}$$

$$2NH_3 + NO_2^- \longrightarrow N_2 + NO_3^- \tag{15.5}$$

Case Study: Enzymatic Removal of Nitrate

As nitrate reduction requires an electron donor such as NAD(P)H, making carbon sources is necessary for its regeneration. A bioelectro-denitrification process was constructed using electricity instead of feeding carbon source as the electron donor (Choi *et al.* 2006; Kim *et al.* 2007). In this process, permeabilized *Ochrobactrum anthropi* SY509 was employed for the denitrification reaction owing to the presence of denitrifying enzymes. When growing the microorganism, the synthesis of the enzymes involved should be induced for maximized synthesis. Electrons could be transferred to the enzymes from an electrode via mediators. Thus, the treatment could increase the accessibility of the nitrate and mediators to enzymes, thereby improving the nitrate removal efficiency. Recently, the process employing three-dimensional

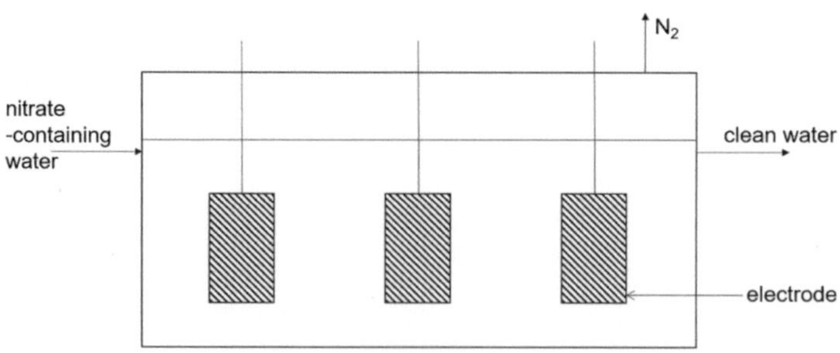

Fig. 15.1 Schematic diagram of nitrate-containing water treatment system employing three-dimensional bioelectrode

bioeletrode which does not require the mediator was reported (Cho *et al.* 2008). Powder from electrode material was mixed with the permeabilized cells which contained the enzymes for denitrification and then were polymerized using conducting polymer material for three-dimensional electrode. Further researches including further development of the technology and scale-up are required (Fig. 15.1).

15.2 Carbon Dioxide Conversion

Carbon dioxide (CO_2) concentration in the atmosphere has been increasing due to anthropogenic activities, and has become a major contributor to global warming. One solution to the problem on CO_2 emission is carbon capture and storage (CCS). In this approach CO_2 is captured in a power plant and sequestered in suitable geologic forms for long-term storage. This approach can be improved by converting CO_2 into value-added products. The process is called CO_2 capture and utilization (CCU).

Since carbon dioxide has very stable chemical properties, it is necessary to switch the oxidized state of CO_2 to the organic compounds in a reduced state through the reduction reaction in order to produce useful compounds from CO_2. Even though a reducing agent such as hydrogen has been available for the reduction reaction, recently novel method utilizing electric energy and hydrogen ion at the same time was introduced. Recent advances in the renewable energy make electrical reducing power increased interest for the reduction of carbon dioxide. In particular, using the electric energy produced from renewable resources energy to convert carbon dioxide into useful compounds has additional advantage in which the electrical energy can be stored in the compound derived from the carbon in addition to useful compounds. However, it is desirable to decrease the required electric energy through the co-use of hydrogen ion (Schneider *et al.* 2012).

Figure 15.2 shows the value of the compound produced and the invested energy for the compound from CO_2. Given the reduction of energy and administered prices at the same time, formic acid and carbon monoxide are deemed useful as priority targets.

Despite of a large global demand for methanol, the price of the compound is expected to be rather low, and less economically efficient considering six electrons required for the electrochemical conversion of carbon dioxide. Finding efficient conversion catalyst is the universal challenge in the electrochemical conversion of carbon dioxide to produce useful chemical products such as formic acid.

15.2.1 Carbon Dioxide to Formic Acid

Enzymatic catalyst to convert the carbon dioxide into formic acid has so far not been reported. However, there is a formate dehydrogenase mainly catalyzing the

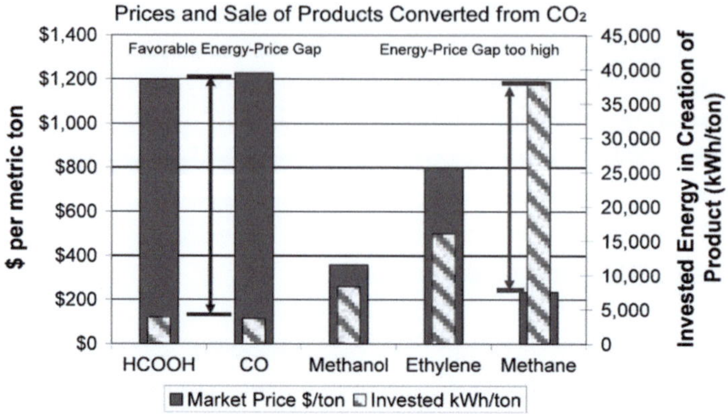

Fig. 15.2 Selection of most suitable target compound considering prices of products converted from CO_2 and invested energy (DNV Report 2011)

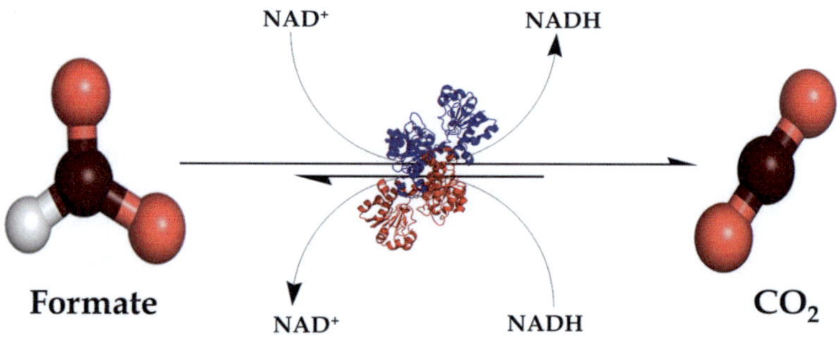

Fig. 15.3 CO_2 conversion scheme by NADH-dependent formate dehydrogenase

oxidation to carbon dioxide and attempts to produce formic acid from CO_2 with the reversibility of this enzyme reaction have been actively tried. Formate dehydrogenases present in the natural world are largely divided into NADH-dependent and independent one. In the case of NADH-dependent formate dehydrogenase, active site generally does not contain a separate metal causing oxygen sensitivity. Carbon dioxide reduction scheme by NADH-dependent formate dehydrogenase can be expressed in Fig. 15.3.

Currently, the biggest hurdle of formate formation is the gap between oxidation rate of formate and reduction rate of CO_2 since the oxidation of formic acid is known much dominant compared with reduction of CO_2. The fact that it is almost impossible to obtain high concentration of formate with known formate dehydrogenase give lesson that discovery of new types of enzyme or protein engineering of enzymes to improve the characteristics are essentially required. Recently, new formate dehydrogenase sourced from *Thiobacillus* showed much

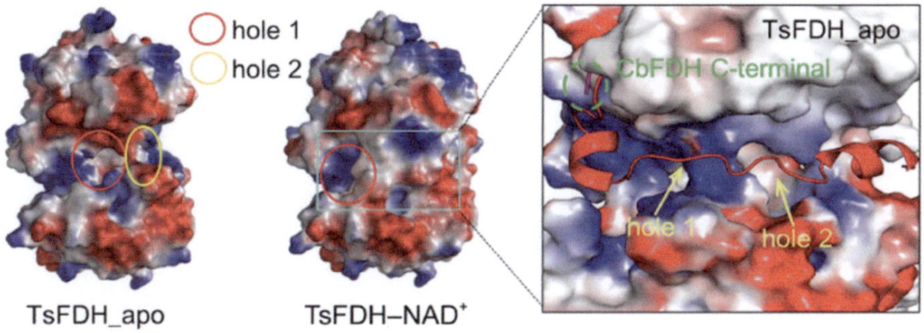

Fig. 15.4 Three-dimensional structure of *Thiobacillus* formate dehydrogenase (Choe *et al.* 2015)

higher performance than the conventional formatted dehydrogenases for conversion of carbon dioxide (Choe *et al.* 2014). The enzyme three-dimensional structure revealed the entrance hole allowing carbon dioxide to transport smoothly (Fig. 15.4) and a relatively low activation energy for reduction of CO_2 to formic acid was suggested by the QM/MM study (Choe *et al.* 2015). But formic acid production rate is not high enough to apply to commercial scale and a further improvement is required. In addition to the problem of enzyme, NADH regeneration is critically required since the reaction requires equivalent amount of NADH for the reduction of CO_2. However, the electrochemical regeneration of NADH cannot be efficiently performed without using another catalyst such as Rh complex.

NADH-independent formate dehydrogenase is known to utilize directly electrons supplied from the electrode without the need for expensive electron carrier such as NADH. This type of enzyme contains metal complex composed of Mo or W pterin ligand at the reaction active site and exhibits the hypothesized reaction mechanism as shown in Fig. 15.5.

These enzymes are capable of performing remarkably fast carbon dioxide reduction rate up to 3400 s^{-1} using electrons supplied from cathode shown in Fig. 15.6 with 95% of Faraday efficiency (Reda *et al.* 2008). However, extremely vulnerable property to oxygen and weak stability prevent the application of this enzyme at large scale.

Almost enzymes present in nature seem more adequately evolved for oxidation to harvest energy from substrate rather than for reduction. Enzymes showing relatively superior performance for reduction reaction were characterized harboring metal atoms at the reaction site which may impose oxygen vulnerability since enzymes have been evolved in anaerobic environment. In order to execute industrially mass-conversion of carbon dioxide improvement of enzyme is expected necessary to increase applicability of enzyme. Based on the protein structure in silico designs actively employing computational chemistry techniques are expected to accelerate the improvement of enzyme for this purpose.

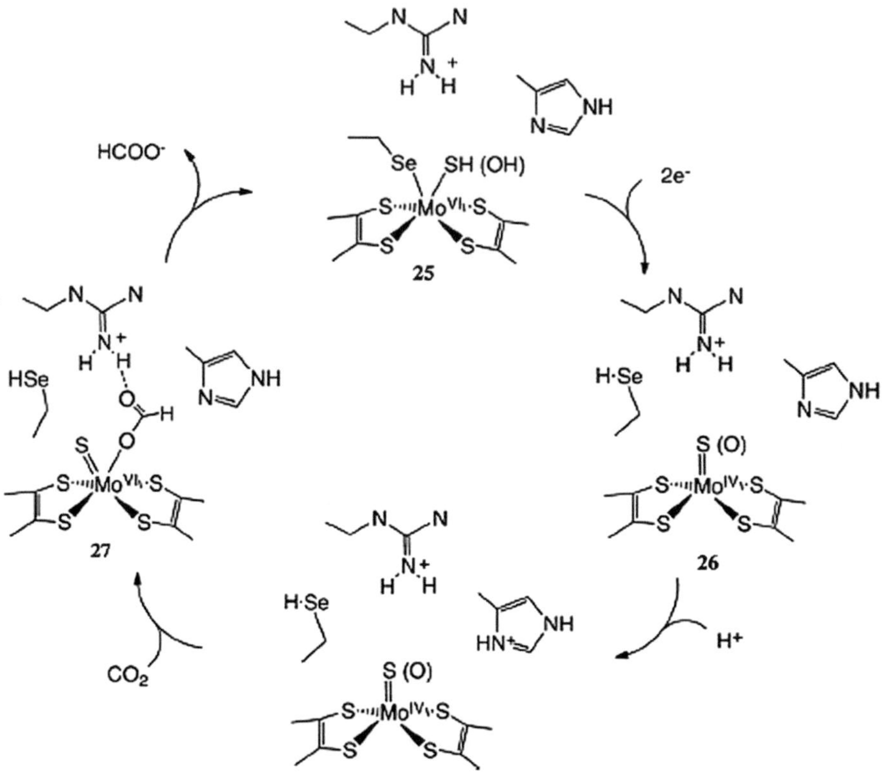

Fig. 15.5 Suggested reaction mechanism of formate dehydrogenase containing Mo-pterin for the reduction of CO_2 to formate (Appel *et al.* 2013)

15.2.2 Carbon Dioxide for Carbonate

Carbonic anhydrase (CA, EC 4.2.1.1) can be employed for CO_2 sequestration. It accelerates CO_2 capture from the extracellular environment into the intracellular system and is very important biocatalyst for hydration or dehydration reactions between CO_2 and bicarbonate (Savile and Lalonde 2011). The steps involved in the indirect CO_2 mineralization are given below:

$$CO_{2(g)} \rightleftharpoons CO_{2(aq)} \tag{15.6}$$

$$\begin{gathered} CO_{2(aq)} + H_2O \underset{k_{-2}}{\overset{k_2}{\rightleftharpoons}} H_2CO_3 \\ K_2 = k_2/k_{-2} = 2.6 \times 10^{-3} \end{gathered} \tag{15.7}$$

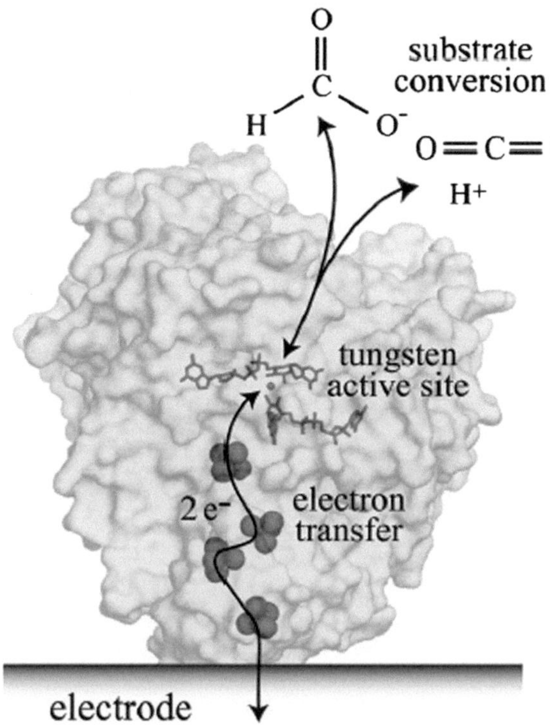

Fig. 15.6 Synthesis of formate from CO_2 using NADH-independent formate dehydrogenase utilizing electrons supplied from cathode (Reda *et al.* 2008)

$$H_2CO_3 \underset{k_{-3}}{\overset{k_3}{\rightleftharpoons}} H^+ + HCO_3^-$$
$$K_2 = k_3/k_{-3} = 1.7 \times 10^{-4} \tag{15.8}$$

$$HCO_3^- \rightleftharpoons H^+ + CO_3^{2-} \tag{15.9}$$

$$Ca^{2+} + CO_3^{2-} \rightarrow CaCO_3 \downarrow \tag{15.10}$$

1. CO_2 is dissolved in water.
2. The aqueous CO_2 reacts with water forming carbonic acid.
3. Carbonic acid ionized bicarbonate and carbonate ions.
4. Calcium carbonate precipitates by association of calcium and carbonate ions.

The rate-limiting step is the second reaction. In this reaction, CA increases CO_2 precipitation rate. Figure 15.7 shows the schematic representation of CO_2 capture

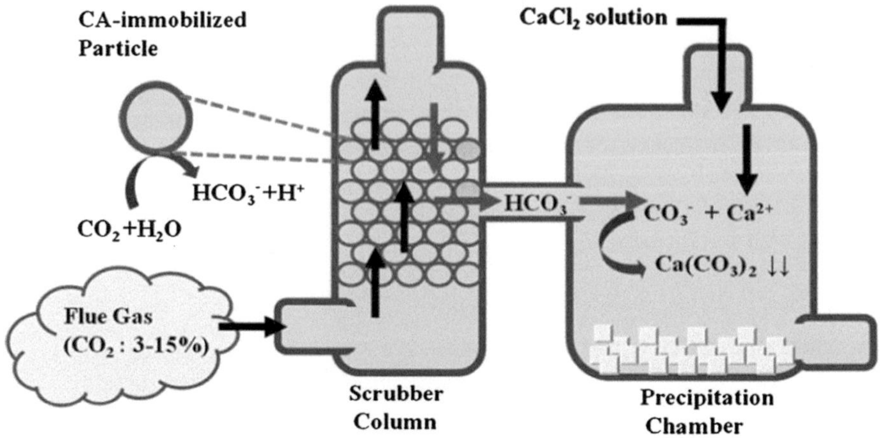

Fig. 15.7 Schematic diagram of carbonic anhydrase-catalyzed CO_2 capture and utilization

and utilization. In this figure, CA is applied as a biocatalyst to a coal-fired power generation plant or other CO_2-intensive industries. CA is immobilized on solid support and used in a scrubber column. Water is sprayed at the top of the column to wash the upgoing flue gas and captures CO_2 from it. The immobilized CA upon contact with the aqueous solution captures and converts CO_2 into bicarbonate ions, which is further converted into calcium carbonate by contact with calcium chloride solution.

Aside from the formation of limestone from CO_2, organic acid such as oxaloacetate (OAA) can be produced using enzymes. Phosphoenolpyruvate carboxylase (PEPCase, EC 4.1.1.31) is the main anaplerotic enzyme, irreversiblely converting phosphoenolpyruvate into four-carbon organic acids, OAA using HCO_3^-. OAA is used for replenishing the intermediates of tricarboxylic acid (TCA) cycle and is a precursor of organic acids such as malic acid and succinate acid. Combining CA and PEPCase for an effective CO_2 capture and utilization system was studied. Recombinant CA and the PEPCase from marine microalgae were expressed in *Escherichia coli* to investigate capturing of CO_2 and converting it into valuable four-carbon compounds in a sequential manner. The PEPCase 1 gene of the marine diatom *Phaeodactylum tricornutum* was cloned and induced to express a recombinant PtPEPCase 1. The purified PtPEPCase 1 had specific enzymatic activity (5.89 U/mg). Higher level of OAA was present when the CA from *Dunaliella* sp. and the PtPEPCase 1 coordinated the reactions than when only PEPCase was available. Recently, engineered enzyme was used for this purpose (Kim *et al.* 2012). However, for practical applications, high enzyme activity at high-temperature and long-term thermostability are to be high enough, which remains for further study (Fig. 15.8).

Two current cases are described as above. However, carbon dioxide reduction is an urgent global issue and thus many technologies are being developed including

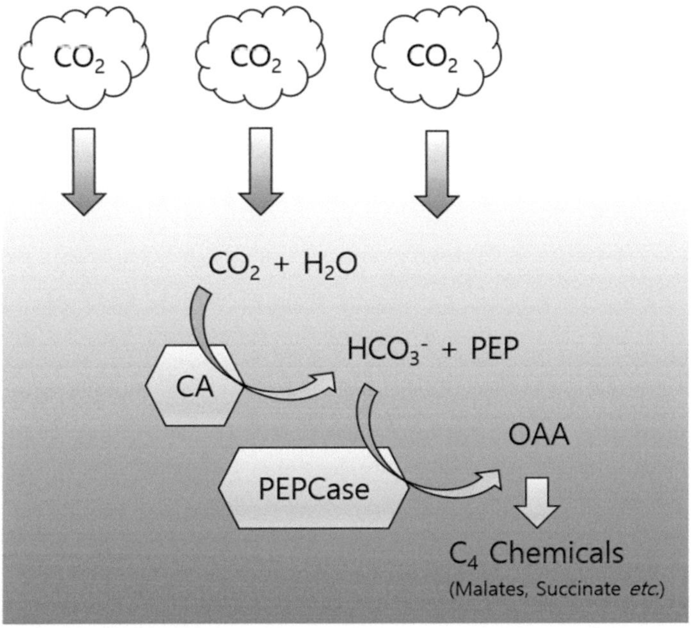

Fig. 15.8 Schematic diagram of carbonic anhydrase and phosphoenolpyruvate carboxylase-catalyzed CO_2 capture and utilization

chemical, physical, and biological methods. Productions of bio-based chemicals and bioenergies from biomass are a good means to reduce carbon dioxide emissions. Algae are also known to use carbon dioxide as a carbon source and thus cultivation of algae for just growth or for useful chemicals production from algae can be a good alternative to reduce carbon dioxide. Another trial was reported to convert carbon dioxide into formate by electrochemical means and then convert it into alcohols (Li *et al.* 2012). Further researches are urgently required to develop useful and efficient technologies on carbon dioxide issue. In any case, enzymatic method should compete with chemical or physical methods, and activity and stability issues are thus remained to be improved.

Further Discussion

1. Mercury ion can be reduced by enzyme, where reduced form is volatile and non-toxic compared to mercury ion. Discuss on the enzyme for mercury detoxification.
2. Many soils and lands are being contaminated by organophosphate pesticides. Discuss how to clean the site by bioremediation.
3. Algae can be used for carbon dioxide removal and for the chemicals production, since algae can utilize carbon dioxide as a carbon source. What technology is to be developed for the commercialization of this technology?

4. Carbon dioxides are being emitted from the incineration after use of the chemicals and polymers of petroleum origin. Can we reduce carbon dioxide emissions using bioenergy and bio-based chemicals?

References

Alemzadeh I and Nejati S. Phenols removal by immobilized horseradish peroxidase. *J. Hazardous Mater.* 2009, 166:1082-1086.

Appel AM, Bercaw JE, Bocarsly AB, Dobbek H, Dubois DL, Dupuis M, Ferry JG, Fujita E, Hille R, Kenis PJA, Kerfeld CA, Morris RH, Peden CHF, Portis AR, Ragsdale SW, Rauchfuss TB, Reek JNH, Seefeldt LC, Thauer RK and Waldrop GL. Frontiers, opportunities, and challenges in biochemical and chemical catalysis of CO_2 fixation. *Chem. Rev*, 2013, 113:6621-6658.

Basheer S, Kut OM, Prenosil JE and Bourne JR. Development of an enzyme membrane reactor for treatment of cyanide-containing wastewaters from the food industry. *Biotechnology and Bioengineering*, 1993, 41(4):465-473.

Caldwell SR *and* Raushel FM. Detoxification of organophosphate pesticides using a nylon based immobilized phosphotriesterase from *Pseudomonas diminuta. Applied Biochemistry and Biotechnology*, 1991, 31(1):59-73.

Choe HJ, Joo JC, Cho DH, Kim MH, Lee SH, Jung KD and Kim YH. Efficient CO_2-reducing activity of NAD-dependent formate dehydrogenase from *Thiobacillus* sp. KNK65MA for formate production from CO_2 gas. *PLOS ONE*, 2014, 9:e103111. doi:10.1371/journal.pone.0103111.

Choe Hj, Ha JM, Joo JC, Kim HO, Yoon HJ, Kim SH, Son SH, Robert M. Gengan, Jeon ST, Rakwoo Chang, Jung KD, Kim YH and Lee HH. Structural and functional characterization of an NAD-dependent formate dehydrogenase from *Thiobacillus sp.* KNK65MA with efficient CO_2-reducing activity, *Acta Crystallographica Section D, 2015,* 71:313-323.

Cho JS, Park JY and Yoo YJ. Novel 3-dimesional bioelectrode for mediatorless bioelectrochemical denitrification. *Biotechnology Letters*, 2008, 30:1617-1620.

Choi KO, Song SH, Kim YH, Park DH and Yoo YJ. Bioelectrochemical denitrification using permeabilized *Ochrobactrum anthropi* SY509. *J. Microbiol. Biotechnol.*, 2006, 16:678-682.

Choudhary RB, Jana AK and Jha MK. Enzyme technology applications in leather processing. *Indian J. Chem. Technol.,* 2004, 11: 659-671.

Dezott M, Innocentini-Mei LH and Duran N. Silica immobilised enzyme catalysed removal of chlorolignins from kraft effluent. *J. Biotechnol.*, 1995, 43:161-167.

DNV Report, "Carbon dioxide utilization" 2011, http://www.dnv.com/binaries/dnv-position_paper_CO2_utilization_tcm4-445820.pdf.

Dumas DP, Durst HD, Landis WG, Raushel FM, and Wild JR. Inactivation of organophosphorus nerve agents by the phosphotriesterase from *Pseudomonas diminuta. Archives of Biochemistry and Biophysics*, 1990, 227:155-159.

Dumestre A, Chone T, Portal J, Gerard M and Berthelin J. Cyanide degradation under alkaline conditions by a strain of *Fusarium solani* isolated from contaminated soils. *Appl. Environ. Microbiol.*, 1997, 63:2729-2734.

Hu BL, Shen LD, Xu XY and Zheng P. Anaerobic ammonium oxidation (anammox) in different natural ecosystems. *Biochem. Soc. Trans.*, 2011, 39:1811-1816.

Hu Z, Lotti T, van Loosdrecht M and Kartal B. Nitrogen removal with the anaerobic ammonium oxidation process. *Biotechnol. Lett.*, 2013, 35:1145-1154.

Jetten MSM, Wagner M, Fuerst J, van Loosdrecht M, Kuenen G and Strous M. Microbiology and application of the anaerobic ammonium oxidation ('anammox') process. *Curr. Opin. Biotechnol.,* 2001, 12:283-288.

Jiang Y, Tang W, Gao J, Zhou L and He Y. Immobilization of horseradish peroxidase in phospholipid-templated titania and its applications in phenolic compounds and dye removal. *Enzyme Microb. Technol.*, 2014, 55:1-6.

Kim IG, Jo BH, Kang DG, Kim CS, Choi YS and Cha HJ. Biomineralization-based conversion of carbon dioxide to calcium carbonate using recombinant carbonic anhydrase. *Chemosphere*, 2012, 87:1091-1096.

Kim SJ, Joo JC, Song BK, Yoo YJ, Kim YH. Engineering a horseradish peroxidase C stable to radical attacks by mutating multiple radical coupling sites. *Biotech. Bioeng.*, 2014, 112:668-676.

Kim YH, Park YJ, Song SH and Yoo YJ. Nitrate removal without carbon source feeding by permeabilized *Ochrobactrum anthropi* SY509 using electrochemical bioreactor. *Enzyme Microb. Technol.*, 2007, 41:663-668.

Klivanov AM and Morris ED. Horseradish peroxidase for the removal of carcinogenic aromatic amines from water. *Enzyme Microbiology and Technology*, 1981, 3:119-122.

Li H, Opgenorth PH, Wernick DG, Rogers S, Wu TY, Higashide W, Malati P, Huo YX, Cho KM and Liao JC. Integrated electromicrobial conversion of carbon dioxide to higher alcohols. *Science,* 2012, 335:1596.

Reda T, Plugge CM, Abram NJ, and Hirst J, Reversible interconversion of carbon dioxide and formate by an electroactive enzyme. *PNAS*, 2008, 105:10654-10658.

Savile CK and Lalonde JJ, Biotechnology for the acceleration of carbon dioxide capture and sequestration. *Current opinion in Biotechnology*, 2011, 22:818-823.

Schneider J, Jia H, Muckerman JT and Fujita E. Thermodynamics and kinetics of CO_2, CO, and H^+ binding to the metal centre of CO_2 reduction catalysts. *Chem. Soc. Rev.*, 2012, 41:2036–2051.

Tatsumi K, Wasa S and Ichikawa H. Removal of chlorophenols from wastewater by immobilized horseradish peroxidase. *Biotechnology and Bioengineering*, 1996, 51:126-130.

Thanikaivelan P, Rao JR, Nair BU and Ramasami T. Progress and recent trends in biotechnological methods for leather processing. *Trends Biotechnol.*, 2004, 22:181-188.

Vasileva N, Godjevargova T, Ivanova D and Gabrovska K. Application of immobilized horseradish peroxidase onto modified acrylonitrile copolymer membrane in removing of phenol from water. *Int. J. Biol. Macromol.*, 2009, 44:190-194.

Zhang Y, An J, Ye W, Yang G, Qian ZG, Chen HF, Cui L and Feng Y. Enhancing the promiscuous phosphotriesterase activity of a thermostable lactonase (GkaP) for the efficient degradation of organophosphate pesticides. *Appl. Environ. Microbiol.,* 2012, 78:6647-6655.

Zheng B, Yu S, Zhang Y, Feng Y and Lou Z. Crystallization and preliminary crystallographic analysis of the phosphotriesterase-like lactonase from *Geobacillus kaustophilus. Acta Crystallogr.Sect. F Struct. Biol. Cryst.Commun.,* 2011, 67:794-796.

Index

Y.J. Yoo *et al.*, *Fundamentals of Enzyme Engineering*,
DOI 10.1007/978-94-024-1026-6

D

E

F

MIX
Papier aus verantwortungsvollen Quellen
Paper from responsible sources
FSC® C105338

If you have any concerns about our products,
you can contact us on
ProductSafety@springernature.com

In case Publisher is established outside the EU,
the EU authorized representative is:
Springer Nature Customer Service Center GmbH
Europaplatz 3, 69115 Heidelberg, Germany

Printed by Libri Plureos GmbH
in Hamburg, Germany